Malta and Gozo Buses

Tom Johnson

&

British Bus Publishing

Body codes used in the Bus Handbook series:

Type:
A	Articulated vehicle
B	Bus, either single-deck or double-deck
BC	Interurban - high-back seated bus
C	Coach
M	Minibus with design capacity of 16 seats or less
N	Low-floor bus (*Niederflur*), either single-deck or double-deck
O	Open-top bus (CO = convertible - PO = partial open-top)

Seating capacity is then shown. For double-decks the upper deck quantity is followed by the lower deck.
Please note that seating capacities shown are generally those provided by the operator. It is common practice, however, for some vehicles to operate at different capacities when on certain duties.

Door position:
C	Centre entrance/exit
D	Dual doorway.
F	Front entrance/exit
R	Rear entrance/exit (no distinction between doored and open)
T	Three or more access points

Equipment:-
T	Toilet	TV	Training vehicle.
M	Mail compartment	RV	Used as tow bus or engineers' vehicle.
L	Lift for wheelchair (post 2005 express coaches are fitted with lifts as standard)		

Allocation:
s	Ancillary vehicle
t	Training bus
u	out of service or strategic reserve; refurbishment or seasonal requirement
w	Vehicle is withdrawn and awaiting disposal.

e.g. - B32/28F is a double-deck bus with thirty-two seats upstairs, twenty-eight down and a front entrance/exit.
N43D is a low-floor bus with two or more doorways.

Re-registrations:
Where a vehicle has gained new index marks the details are listed at the end of each fleet showing the current mark, followed in sequence by those previously carried starting with the original mark.

Annual books are produced for the major groups:
The Stagecoach Bus Handbook
The First Bus Handbook
The Arriva Bus Handbook
The Go-Ahead Bus Handbook
The National Express Coach Handbook (bi-annual)
Some editions for earlier years are available. Please contact the publisher.

Regional books in the series:
The Scottish Bus Handbook
The Welsh Bus Handbook
The Ireland & Islands Bus Handbook
English Bus Handbook: Smaller Groups
English Bus Handbook: Notable Independents
English Bus Handbook: Coaches

Associated series:
The Hong Kong Bus Handbook
The Malta Bus Handbook
The Leyland Lynx Handbook
The Postbus Handbook
The Mailvan Handbook
The Toy & Model Bus Handbook - Volume 1 - Early Diecasts
The Fire Brigade Handbook (fleet list of each local authority fire brigade)
The Police Range Rover Handbook

Some earlier editions of these books are still available. Please contact the publisher on 01952 255669.

CONTENTS

Acknowledgments:

In writing this book, the author has received special help from many people, both at home and on Malta and Gozo. Victor Spiteri, Chairman of the ATP, Malta's Public Transport Association, has provided extensive up-to-date vehicle details. Thomas Knowles has checked and supplemented many of the new details about the islands' vehicles presented in this second edition. Richard Stedall has allowed the author unrestricted access to all his *Rummagings in the Archives'* which he has undertaken personally in recent years and Peter Griffiths has researched the Duple and Plaxton bodywork on the islands' coaches. John Veerkamp confirmed details of the King Long vehicles on Malta, whilst Marthese Camilleri at Naxxar local council offices provided current details of the Naxxar local services. Joe Xerri provided new and up-to-date facts about the Gozitan bus scene. All their help has added to the author's own research. Brendan Fox planned and drew the route maps which must rank as the most precise ever published about the islands' transport layouts. Special thanks must be given to Marco Zammit who has provided first-hand local and historical details, photographs and advice throughout the preparation of the book, and without his help, this *Bus Handbook* would have been the poorer. All photographs are by the author unless shown otherwise.

1st Edition - 2003 ISBN 1-897990-97-9

ISBN 9781904875581

Published by *British Bus Publishing*
16 St Margaret's Drive, Telford, TF1 3PH
© Tom Johnson and British Bus Publishing, February 2010

INTRODUCTION

This second *Malta Bus Handbook* differs in approach from many others previously published by *British Bus Publishing* and is published just prior to the whole-scale changes to be introduced following recommendations of the 2008 Halcrow Report. This new edition therefore presents a picture of what has become the norm on the island over the past ten years or so. Each route bus on Malta and Gozo and each coach is listed, accompanied by full technical details, current ownership and previous history. Other developments are covered, with details of the Open Top vehicles and changes to the Unscheduled Buses fleet (coaches) and minibuses on Malta and the few developments on Gozo. The opportunity had been taken, also, to include photographs and vehicle details of the islands' minibuses, the Police bus fleet, the Educational fleet, Airport vehicles and some withdrawn wrecks on Malta. Chapters cover the bus routes on both Malta and Gozo, with route descriptions, details of sights to see and places to visit en route, also include a photograph of a vehicle in service somewhere along the route. Indeed, it has been the author's aim throughout to include photographs of vehicles in service around the island, rather than limiting himself to the all-too-easy task of photographing an abundance of vehicles at City Gate bus terminus around the Triton Fountain outside the entrance to the capital city. Any bus enthusiast who has yet to visit Malta and Gozo will, it is hoped, be encouraged to pay a visit soon, if this series of photographs succeeds in its aim of presenting vehicles in service in picturesque, historical or architecturally interesting settings. Any tourist attracted to the buses by this book will find out about the bus he is travelling on as well as details about the islands.

DBY 460 is a Bedford SB which carries a very distinctive body built by Brincat with a unique style of destination board and a passenger door which is an unusual feature for locally built vehicles. It is parked at the Sliema Ferries terminus of route 652 to Golden Bay one morning.

A SHORT HISTORY OF PUBLIC TRANSPORT ON THE ISLANDS

The Maltese archipelago is situated some 93 kms south of Sicily in the Mediterranean Sea, about 1,826 kms from Gibraltar to the west and around 1,510 kms east of Alexandria. There are three inhabited islands, Malta, Gozo and Comino, with the other smaller uninhabited ones of Cominotto, St Paul's Islands, Manoel Island and Filfla, which is now a nature reserve and has been a bird sanctuary since 1988. There are also several other minor uninhabited rocks most of which are home to rare and endemic species of their own. About 414,000 people live on Malta in an area of 246 square kms. There are roughly 31,000 residents on Gozo, which is about one third the size of Malta, and there are just eight permanent residents on Comino. Malta is divided into 54 localities and Gozo, including Comino, is divided into fourteen. Since 1993 each local council has been elected for a three-year term with one third of the councils being elected during each annual election, which is usually but not always held in March.

Despite its small size, Malta can claim three forms of mechanical land transport systems throughout its history. Initially a railway between Valletta and Mdina (then still referred to as Notabile) near Rabat, was opened in 1883. This metre-gauge system covered about eleven lilometres from the capital to the 'old' capital, passing through the most densely populated areas of the island with stations in Hamrun, Birkirkara and Attard. Never a commercial success, the business went bust some years later and the Government intervened, reconditioned the track and restored services in 1892. Soon afterwards work began on extending the railway to Museum on the far side of the hill on which Rabat

Volvo Saracakis EBY 598 is owned by Philip Caruana and is stopped in Lija for this photograph which shows the beautiful surroundings of one of the Three Villages in September 2009.

The remains of withdrawn vehicles can be found in many fields and garages throughout the islands. This Thames normal control bus with Schembri bodywork can still be found in a field in the north-west of Malta, despite having been withdrawn from active service about twenty years ago. It is the former 1891, which later became Y-0413. *T W W Knowles*

stands. This was to serve better the barracks at Mtarfa. A 700m extension that included a 550m tunnel was constructed and new locomotives were purchased and delivered in 1905. Journey times were approximately thirty minutes, but competition from the tram services introduced on New Year's Day 1906 immediately threatened the viability of the railway, especially as one of the tram routes served Birkirkara, in direct competition with the trains. The electric tram operation boasted both toast-track and conventional trams on three routes from Valletta. The tramways were run by Macartney and Mc Elroy Company, which also operated the Barracca Lifts, which opened in 1903.

The railway and tramway systems were not all, for in 1904 the Malta Motor Bus Company was instituted and imported a Thornycroft bus to run between Valletta and St Julian's. This proved an immediate success and so more buses and a lorry were purchased. These included 36-seat double-deck buses that were delivered in 1905, which carried a khaki livery with black lining. Each had a destination board on each side, as well as the registration number below. The three forms of transport entered into direct competition, yet were very little affected by the First World War. Thereafter numerous small bus firms started, mainly one-man concerns operating as they wished showing little acknowledgment to any regulations. The British Motor Company was set up in 1920 and by 1922 was providing a regular service between Valletta and Sliema. Its initial fleet consisted of a couple of lorries and six Daimler ambulances converted into buses. The company immediately provided fierce competition with the Marsamxett ferries which operated between Sliema and Valletta. East of Valletta, the Cottonera Motor Car Company was formed in 1921 and eventually went on to compete with the Grand Harbour ferries. Fifty bus licences had been granted by the end of 1922 and five years later their number had risen to 120. These licences refer to the registration of each vehicle, and as a bus is bought

Malta and Gozo Buses

and sold, its registration number is transferred, even to this dayThe British Motor Company established a better fleet of buses during these early years, but in 1929 a British-owned company, the Overseas Motor Transport, controlled by a Commander Hare of Devon Motor Transport, looked abroad to expand its business and decided to buy the British Motor Company Limited of Malta. The OMC retained the BMC name and set out to replace its acquired mixed fleet of small old buses with around thirty much better Thornycroft BC buses with Hall Lewis bodies. These vehicles operated services from the Porta Reale terminus to the heavily populated areas around Sliema, Kalafrana via Birzebbugia and Cospicua which competed directly with one of the tram routes. The company had connections with ferry and seaplane services too. Also, during 1929, the tram routes were withdrawn and the Rabat Service Company commenced morning bus services and this spelt the beginning of the end for the railway operations. Two years later, on March 31st 1931, the railway was closed down and the smart dark blue Rabat buses of the RSC took over services the following day.

1929 also saw the introduction of the first amendments to the Maltese 'Motor Transport Regulations' to include public service vehicles. The following year the constitution was suspended; the island's Governor called for a committee to report on the future development of the motor bus service. The newly created Traffic Control Board of 1931 became responsible for new rules governing the buses; bus operation was consolidated into eleven routes, each identified by its own colour scheme and therefore easily recognisable by passengers, many of whom, in those days, were very poorly educated and could neither read nor write. Licences were issued allowing buses to operate specific routes, only. Bus-building guidelines came into effect and then some routes running about fifty buses were re-located to the newly opened Piazza Regina terminus. However the route situation remained rather fluid and as early as 1933 some routes had already split resulting into smaller groups and in some cases, other livery variations.

The BMC still found bus operation difficult, with continued competition from the rival Sliema Motor Bus Service and other family concerns and so its British owners decided to sell their Maltese interests to Joseph Gasan in 1933. The BMC along with its fleet of Thornycroft buses went on to work in Cairo. Gasan had started way back in 1919 with ship recovery and ferry services. His most successful move was in 1928 when he became the local Ford agent. Gasan concentrated his bus services mainly on the Sliema area and the route between Birkirkara and St. Julian's. Gasan expanded his BMC fleet which was predominantly on Ford chassis. His vehicles carried fleet numbers, an unusual feature in the history of Malta's buses. These were very likely inherited, again along with the BMC name, from the previous concern. Gasan however adopted his own, rather austere dark green livery that was broken only by a bold white stripe along the waist-line for his buses. The deal also included the Gozo Mail bus service. His premises in Gzira boasted well-equipped workshops to deal with repairs and overhauls of his fleet. Many complete buses were built in-house, nearly all on Ford chassis of primarily American origin.

By the beginning of the Second World War the former separate 'route managers' had been ousted and the bus operations started to be run jointly between the Police and the TCB from October 1940. Unrest followed November 1941, when bus licences were increased to £6 sterling on Malta. To make matters worse, rationing and lack of serviceable vehicles forced the Army garrison to requisition buses as well from 1942 onwards. One AEC S-type open-top bus from Gasan's BMC fleet was used by the RAF as a fuel tender, and fighter aircraft were also overhauled in his facilities in Gzira. Oil and petrol consumption was paid for by the Army, but the costs of mechanical repairs were deducted from the dividends to each respective owner. Several buses were camouflaged and some had their backs cut open and most seats removed to convert them into removal vans and rudimentary emergency vehicles. Even old rear-entrance buses which by then had been off the road for

It was sad but inevitable that so many of the historic route buses were withdrawn and eventually disposed of during the mid-1990s, when the low-floor vehicles arrived as their replacements. It is encouraging to know, however, that other withdrawn vehicles are lovingly stored away in various garages and well-protected sheds on Malta, especially. This view shows from left to right the former DBY 313, a Ford ET7 with Casha bodywork of 1953 vintage (the original 159 fleet number allocated to this licence is just visible), Y-0462, a Bedford OB with Aquilina B36C bodywork from 1950 vintage and DBY 731, a Ford ET7 also with Aquilina bodywork, again from 1953. *TWJ/Marco Zammit*

more than a decade were pressed into service o again. Provision was made for owners of camouflaged buses to claim their repainting expenses following the end of hostilities. In 1943 an attempt was made to amalgamate the bus operations, but it failed.

Following the Second World War and its consequent problems on the war-torn island, Gasan and all other operators tried to rebuild their respective bus businesses as soon as possible. Suppliers in Great Britain were unable to be of much help, as rebuilding was the order of the day there too, at that time. In spite of this, the resourceful Gasan Enterprises were able to complete their first new post-war bus by March 1946. This was a Fordson with a Gasan 34-seat centre entrance body, registered as the 'new' 2543. This was initially placed in service on the Balluta Bay service and carried fleet number 22 in Gasan's post-war revised fleet number series. However service demands were still exceeding the available supplies and in order to continue, Gasan purchased bus body kits from Wayne Bus Builders and Suppliers of Richmond USA. With more buses in service Gasan increased his portfolio to well over one hundred vehicles. It was not until a couple of years later that Gasan was able to buy any brand new chassis from the UK to renew the remaining older vehicles in his fleet. So, some Thames chassis were imported and were fitted with bodies built by other local builders. To many Maltese, these Thames were to become Malta's iconic normal-control buses during the following decades. Gasan's influence on and involvement with the Maltese bus scene came to an end in the mid-1950s, when he sold his bus interests to individual owners on the island. A typical new Thames normal-control bus changed hands at £4,600, complete with its registration.

After the war the termini were renamed to Kingsgate, instead of Porta Reale, and Castille Palace, instead of Piazza Regina, so English names replaced the former Italian ones. Well into the 1950s fine tuning of the routes led to several changes and bus transfers as required from time to time. The sixteen route colours during the 1950s were as follows:

1) Routes departing from the Kingsgate Terminus

Birkirkara and Hamrun	98 buses	bright red
Cospicua/Senglea/Vittoriosa	80 buses	spray green
Rabat/Dingli/Mtarfa	42 buses	dark blue
Siggiewi	5 buses	light yellow with dark red waistband
Zebbug/Qormi	18 buses	golden ochre

(*These two routes were merged in the mid-1960s and adopted a yellow livery with an orange band, one that was very similar to the present-day route livery*)

Sliema/St Julian's/St Andrew's	104 buses	mid-green with dark green band
Zabbar/Marsascala	15 buses	red with a mid-blue waistband
Zurrieq/Luqa/Mqabba/Qrendi	18 buses	bright orange

2) Routes departing from the Castille Terminus

Gudja/Ghaxaq	5 buses	light blue with a cream waistbanL
ija/Attard/Balzan	12 buses	red with a white band
Birzebbugia/Kalafrana/Hal Far	27 buses	light blue with bright red and white waistbands
Mellieha/Mgarr	34 buses	white with a mid-blue waistband
Mosta	18 buses	chocolate brown
Naxxar/Gharghur	9 buses	golden brown
Zejtun/Marsaxlokk	16 buses	deep red with green waistband
Also		
St Julian's to Birkirkara	4 buses	mid-green with a red waistband

During the 1960s some groups merged paving the way to eventual amalgamation. In 1973 the first sign of radical change came with the merger of the Route Buses into three groups, each identified by its own livery. Group A adopted the spray-green livery and served the south and east of Malta. Group B adopted the white livery with a blue band for the Sliema routes and those to the north of the island. Group C adopted a red livery, serving the centre and south-west of the island. This rationalization can be seen as an intermediate change; bus owners had to admit that such changes were required, if only for them to be able to charge higher fares. Several operators even wanted to adopt the former Mellieha livery when the final amalgamation came along. Within two years the 1975 legislation brought further changes. The Castille Palace terminus was finally abolished and all routes from then onwards were operated from City Gate. Finally the government had the last word and all buses were painted in the spray-green livery of the former Cospicua routes. In November 1977 Malta Bus Services Ltd was created to take over the administration of bus operations, but the Police were still responsible for the licensing, testing and also the upkeep of the vehicle files. One-man operation was introduced in 1979.

The 1980s were marked by the importation from Great Britain of numerous buses, such as AEC Swifts from London Transport, Bristol LHs and Tiger Cubs. Several of these were first used during a transport strike in mid-1981. Some were sold for the sum of 4,600 Malta liri each, without registrations, to a number of operators who were about to scrap or were being forced to replace their older vehicles. Numerous Bedford and Ford coaches with

For a while in the early 2000s, route 427 from Qawra to Marsaxlokk served the small residential estate of Tad-Dib on the outskirts of Mosta. Y-0783, a Bedford OL with Casha B36C bodywork, rested here for a few months and was photographed in August 2001.

Plaxton and Duple bodies were imported next. Nearly all had sealed side-windows and so they were very uncomfortable during the hot summers. Many of these imports are still providing yeoman service on the island to this day but by the mid-1990s they had been fitted with sliding side-windows. Many have bus seating nowadays and some have even had their roof-racks removed to make them more passenger-friendly.

In 1989 an Act of Parliament created the Passenger Transport Association which remains as the current scheduled bus operator. The green livery lasted until the mid-1990s, when it was replaced by the current yellow and orange livery, which evokes memories Qormi-Zebbug-Siggiewi group livery of the late 1960s. The current registrations coincided with the arrival of the first five low-floor imports during 1997. Automatic ticket machines were introduced next and these remain in everyday use. About 130 more buses were replaced during 2003 by brand-new low-floor vehicles. Ten more replacements were sanctioned by the ADT during 2007. This time however, the imported buses were a mix of second-hand Leyland Lynx and Dennis Lances with Wright bodies, as well as a couple of Alexander-bodied Volvo B6 buses. A further Leyland Lynx was imported early in 2008 with the intention to replace yet another withdrawn bus, but permission was denied.

During 2009, Cancu Supreme Travel introduced the first two coaches and in January 2010 three brand-new open-top buses of the latest Euro 5 series entered service. Lepeirks Travel of Gozo purchases a King Long XMQ6900 Euro 5 midicoach in late 2009. All the coaches are new demonstrator vehicles.

VEHICLES ON MALTA AND GOZO

The bus station outside City Gate on the edge of Valletta remains the main centre of bus operation. The majority of the routes on Malta originate or terminate at this bus terminus which is overlooked by the Triton Fountain. These routes serve the main localities throughout the island. 'Direct' services provide faster and more convenient links between some of the principal holiday resorts and tourist areas, especially from Bugibba, Qawra and Sliema, thus omitting Valletta altogether, and thereby obviating a change of bus en route. In the more recent past, other direct routes from villages to the Mater Dei Hospital complex and the neighbouring University of Malta have been established on an hourly schedule. Other operations include the Open-top services and the Three Cities Tour from Sliema Ferries. Two other services, the Malta Bus tours of the mid-2000s and the recently withdrawn Gozo Airport Shuttle between Malta International Airport at Luqa and the Gozo ferry terminal at Cirkewwa are also worth noting.

There are 508 yellow and orange route buses on Malta, which are allocated registration numbers from DBY 300 to 470, EBY 471 to 640 and FBY 641 to 807. All the buses are privately owned and many are driven by their owners according to their allotted 'day in and day out' day roster, or on occasions nearly every day. The variety of buses in this fleet has long been Malta's attraction to the bus enthusiast. A few normal-control vehicles, now over fifty years old, remain in operation, along with slightly more modern vehicles with locally-built bodywork by firms such as Aquilina, Barbara, Brincat, Casha, Debono, and Farrugia. Some of these bodywork firms are even today, forty years or more since they closed down, still well-represented but others far less so, with some, such as Daina, Xuereb and Gauci, being represented by just one or two vehicles carrying their bodywork.

Since the first edition of this *Bus Handbook*, some of the long-established veterans have been withdrawn and have been replaced by modern low-floor vehicles originating in China, Eastern European countries and Turkey. The most recent arrivals are second-hand low-floor vehicles from Britain. Enthusiasts from Great Britain will recognise these and many other vehicles which have been imported into Malta, after years of service in England and Wales, especially. Service buses from well-remembered British fleets, such as Crosville, Western National, Southern National and Hants and Dorset are represented by Bristol LH buses. Vehicles from municipal fleets, such as Blackburn, Hyndburn and Cynon Valley, as well as many Bedford and Ford coaches from independent companies all over Great Britain are well represented. AEC Swifts from London Transport have been on the island for thirty years, as the earliest of the batch was imported in 1980.

New low-floor vehicles first appeared in Malta in 1997, when five 45-seater buses were delivered, an Optare Excel and four Dennis Darts. In July 2001 the first of the new régime of foreign vehicles arrived, a Chinese King Long low-floor bus. This vehicle marked the start of a planned renewal of the fleet with new low-floor buses. The first Malta-built low-floor bus, a Volvo B6 with locally-built Scarnif bodywork was the precursor to eight similar MAN 14-220 vehicles and one Volvo B7.

The provision and purchase of new vehicles for the Maltese owners had been a long and protracted affair. Back in 1995 almost 150 owners each paid a deposit of 500 Malta liri (Malta's pre-Euro currency) towards the cost of their new bus. An agreement about the financing of the purchase of new buses was finally concluded just before Christmas 2001. The owners who paid the deposits received a government subsidy of Lm 32,000 for each bus. The balance, which varied between Lm 13,000 and Lm 20,000 depending upon the make and model of the bus which the owner chose, had to be paid by the owner. The new buses had to comply with certain specifications laid down by the Authority. The low-floor buses had to provide easy access for all, including senior citizens, the physically disabled and parents with prams. Engine exhaust emissions had to meet the stringent Euro 3 conditions.

A typical day's work for a Malta bus driver lasts normally about sixteen hours. The first timed departure on many of the routes is as early as 0530 and buses operate, generally, on the same route throughout the day until 2200. The drivers are committed to operate according to a timetable provided by the PTA. This timetable is issued every fortnight and indicates which shift on which route each bus will operate each day during that two-week period. On their days-off, drivers are allowed to use their buses on unscheduled work for instance, in term-time, school journeys, to undertake private work or to carry out repairs. The number of bus departures throughout Malta on any day of the week averages about 4,700, with more during the peak summer months. Consequently on a summer's day, especially during June, July and August, the total number of vehicles in operation may exceed the average of 300 normally in service. A network of night services has been widely advertised recently, too. The Paceville area is noted for its discothèques and night-time entertainment and so there is a regular bus service from here to the principal villages on the island during the early hours at weekends.

The scheduled routes are designated merely by route numbers. Destinations are nowadays only very rarely displayed, although many of the buses of British origin retain their destination boards. Vehicles carry a three-rack number box or board, which is illuminated at night, in the front windscreen, generally. Since February 2008, the route number has also had to be shown in the rear window.

The route numbering system unites routes into areas; those routes passing through Marsa are numbered between 1 and 39; those passing through Msida from 40 to 69 with the Sliema routes numbered in the 60s, and those passing through Hamrun 70 to 94. All these routes depart from Valletta City Gate bus terminus. Route variations numbered above 100 reflect their partner route below 100. For instance, routes 38 and 138 serving Hagar Qim, or routes 15 and 115 which serve Santa Lucija. Special tickets are also available for frequent travellers; 1-day, 3-day, 5-day and 7-day tickets entitle the holder to unlimited travel on the island's bus routes. Otherwise tickets are purchased from the driver of the bus, which in the height of summer can be a lengthy and time-consuming process on the popular tourist routes.

Some of the traditional, older Maltese routes buses display a Madonna and Child in a glass case in the framework surrounding the driver's cab. The colourful decoration and artwork is called *tberfil* and only a very small number of skilled local artists tackle this work.

EBY 506, A Ford V8 forward control vehicle with Casha bodywork swings round one of the hairpin bends as it heads out of Mellieha down to the coast at Ghadira Bay. It is working route 48 from Qawra to Cirkewwa in September 2009. It is interesting to compare the appearance of the vehicle nowadays with its 2001 livery with reference to the photograph on page 22 of the first edition of the *Malta Bus Handbook*.

There are about 430 bus-owners on Malta who have their own collective organisation called the Public Transport Association which was established by Act No. IX of 1989. Its remit is to 'provide, regulate and promote an efficient, economical and secure public transport system on both land and sea'. It has the task of overseeing the daily management of bus operation on the island, collecting the revenue, the profits of which are then shared out amongst the bus owners every fourteen days. In recent months it has committed itself to provide more information about the bus services on the island. In all bus termini in Malta, and in Victoria, Mgarr, Marsalforn and Xlendi on Gozo, notice boards have been erected displaying departure times of all services from that locality. At the Porte des Bombes and at Blata l-Bajda similar information is available about the all the routes serving those two locations.

The PTA is run by a committee of seven chosen from amongst the ranks of the 430 bus owners, and elected by the bus owners. The committee members serve for two years. The committee regularly meets the police and local councils to discuss mutual concerns. A board of discipline meets every fifteen days to oversee reports made by ticket inspectors and passengers. Contentious issues are formally discussed in the presence of an independent lawyer, so that as many as possible are resolved without court actionIn addition to the scheduled buses (the route buses) there are the fleets of some 155 coaches, more than 400 red mini-buses, 200 taxis, and 82 horse-drawn cabs which are called *karrozzini*. Many of the minibus owners belong to the Minibus Co-operative and some of the taxi-drivers are members of the White Taxi Service Amalgamated. Likewise, a number of coaches operate within the KopTaCo Coaches Co-operative. Nowadays the coaches boast a wide variety of livery styles and colours. Many of them are to be seen conveying

FBY 022 and FBY 027 are two of the venerable buses on Gozo. FBY 022, on the left, has been in service for over forty-five years, and FBY 027, on the right, worked in Malta originally. Seen parked in front of the parish church in Nadur on 26th March 2009, the vehicles had been hired for an enthusiasts' tour that day. Notice that the two clock-faces show different times so as to confuse the devil.

holiday-makers to and from Malta International Airport at Luqa at the start and end of their holidays, or conveying them around the island on day-trips.

On Gozo there is now a fleet of seventy-eight vehicles which comprises both service buses and more modern coaches. The fleet is operated by the Gozo Bus Owners' Association. There was a twofold increase in the number of vehicles on the island from the mid-1990s. It thus confirms the increase in popularity of Gozo as a tourist destination. All the daily bus services depart from or near to the bus station in the island's capital, Victoria. In many cases an individual vehicle may be enough to provide the service one route for any given day, especially during off peak hours. However, the service to Mgarr Harbour which is the terminus for the ferry service to Cirkewwa on the north-western tip of Malta is always more demanding especially during the high tourist season. Coaches and buses all carry the same grey and red livery with an off-white roof, though the application of the red band or bands tends to vary from bus to bus. Several nowadays also feature different shades of grey 'relief' making the liveries on these look even more striking. Most of the newest coaches on Gozo are specially imported modern chassis with Maltese-built bodies by Scarnif. This is an anagram of 'Francis' the forename of the owner of the factory; Francis Attard. The coaches cater for the crowds of day-visitors to Gozo, many of whom are on excursions arranged by the tour operators at the hotels on Malta. Buses on Gozo, as in Malta, are owned by individual drivers, or, increasingly, by a firm or family which may have as many ten vehicles. All seventy-eight vehicles are numbered sequentially, without separate numbering schemes for route buses and coaches, as is the case on Malta.

The newly set up Malta Transport Authority (MTA) is to oversee all matters of land, maritime and aviation transport as well as roads infrastructure and traffic management. This has replaced and has more legislative powers than the former ADT.

THE ISLANDS' REGISTRATION SYSTEM

The history of Maltese registrations dates back to 1916, when the original system was introduced and which lasted until 1952. Registration plates were black with a serial number in silver or white, preceded by an eight-pointed Maltese Cross. By 1933, every fifth vehicle registered in Malta was a bus, and most vehicles still bore registrations which were wholly numerical. However, some old photographs show certain buses carrying a prefix or suffix letter during the twenties and thirties. Whilst this first registration series was current, vehicles were allocated registration numbers in consecutive order, irrespective of vehicle type. However, when a newer vehicle replaced an older one, and this was especially true with bus owners; the original registration was retained. Indeed, many of the owners became so fond of these original registrations that many of the vehicles in the current fleet still bear this original number hand-painted usually on the front. Following the end of the Second World War a red disc was included on the right-hand side of the number-plate of vehicles which had permission to enter Valletta, a requirement which was discontinued in 1959.

In 1952 the new scheme consisted of a simple serial number with a plate colour indicating the class of vehicle. Buses, along with most vehicle types, carried white numbers on a black plate. Minibuses for hire, however, bore white plates with a red prefix KIRI followed by two numbers. There was a mixed fleet twenty five midi-coaches and mini-buses in this series. Unlike other mini-buses these were able to work the airport transfers. The international oval for Malta was originally GBY, but this identity was withdrawn in 1967, when M was introduced.

The next major change came on August 1st 1979, when "German-style" plates were ushered in. Measuring 410 x 120mm they featured a white background with a raised black print which consisted of one letter, followed by a hyphen and a serial number of up to four digits. To the right of these numbers was the letter M in a circle. Then buses carried an A prefix before their existing registration number. Thus, the bus which hitherto carried registration 218 became A-0218 in the new scheme. However there were exceptions to this rule, such as a quartet of Gozo buses allocated a five-number registration. Instead of the A prefix and four numbers, three of this quartet eventually received a registration in which the first of the five numerals was replaced by the letter C.

For two and half years, this system ran smoothly, until January 1st 1982, when the A series was allocated to private cars instead, and consequently, buses and coaches received their new red Y block registrations, starting at 1001. Three years later a further change saw buses and coaches adopt a second red series number plates, still within the Y system. However, the opportunity was taken to organise the registrations within this Y series into specific sections:

Y-0001 to Y-0200 for taxis in Malta (originally allocated up to Y-0299)
Y-0201 to Y-0250 for taxis on Gozo (originally Y-0950 to Y-0999)
Y-0300 to Y-0807 for buses on Malta
Y-0808 to Y-0835 for buses and coaches on Gozo (later as far as Y-0846)
Y-0850 to Y-0907 for coaches on Malta (later the allocation was extended to Y-0950)
Y-1000 to Y-1599 for black hire cars and cream chauffeur-driver minibuses
Y-2001 to Y-2400 for red minibuses at first seating up to twelve passengers. To begin with only York-engined Ford Transits were allowed.

Later in 1995, however, on November 5th, this system was scrapped and the current series commenced, with the intention that all existing plates should be replaced within twelve months. Even as late as July 1998, a number of the Maltese route buses still carried their 'old' red plates, and a few still carried the red plates at the start of the following year. Even though Malta did not become a member of the European Union until January 1st 2004, the new registration system called for the inclusion of the blue Euro band with the white letter M beneath a circlet of twelve gold stars. The plates are normally the standard European 520 x 110mm in size and all have a hologram of 'Republic of Malta' crest above a unique seven digit serial, and the substrate is 'watermarked' with a Maltese cross. The plates have a white reflective background with raised characters in black 'FE-script' of the sans-serif series. Registrations consist of three letters and three numerals, the first letter denoting the month of issue; A, M, Y January, B, N, Z February, C, O March, D, P April, E, Q May, F, R June, G, S July, H, T August, I, U September, J, V October, K, W November and L, X December.

From May 2009, a new 'post regulation' registration system for buses and coaches was introduced and the first Unscheduled buses and minibuses bearing this xPY nnn scheme are now already in service on both islands. This new registration system will eventually come into effect on all the PSVs on the Maltese islands.

The 508 route buses on Malta are numbered consecutively within three blocks; DBY 300 to DBY 470, EBY 471 to EBY 640 and FBY 641 to FBY 807. Notice how the route buses merely transferred their numerical registration from the Y-0300 system to the current DBY 300 etc format.

The vehicles on Gozo now have their own new allocation, FBY 001 to FBY 079, with 069 omitted. But this has not always been the case. At the time of the previous Y series, there were only thirty-nine buses on the island, and they bore registrations which followed directly on from the Maltese system, i.e. Y-0808 to Y-0846. When the present system came into force, Gozitan buses did not merely adopt their corresponding FBY mark, as happened on Malta. Instead a completely new system was applied, with an almost coincidental Governmental decision to allow bus owners to buy new vehicles, in an effort to promote tourism and raise the profile of the island. Thus there are now seventy-eight buses and coaches on Gozo. Within the series the numbers can be broadly sub-divided as follows: ex-demonstrator vehicles carry the lowest numbers, FBY 001 to 006; the 39 'original' vehicles are registered between FBY 015 and 057, but there is no correlation between the order of the former 0808 to 0846 sequence and the current one. Newer coaches are primarily allocated higher numbers up to 079. There are a few interlopers within these broad divisions filling in the gaps between 007 and 014, especially.

Tour coaches (Unscheduled buses) on Malta carry xCY registrations, starting with ACY 850 and continuing to LCY 992, and then LCY 001 and CCY 002. Within this sequence the numbers 926, 927 and 928 were not allocated and instead the three 'old timers' which would have assumed these numbers came out of sequence, having adopted registrations LCY 002, LCY 003 and LCY 004 instead. Only ACY, BCY, CCY, JCY, KCY and LCY letter combinations have been issued.

The red minibuses vehicles carry registrations in sequence, and as with the Route Buses their registrations are divided into three almost equal blocks; GMY 001 to GMY 130, HMY 131 to HMY 262 and IMY 263 to IMY 404. In addition, a few different xMY registrations have been allocated at the request of the vehicle owner, so that, for instance, FMY 001, LMY 002 and LMY 007 are also to be seen. Up to around five years ago this minibus fleet consisted entirely of various Ford Transit vans, but there is now a greater variety of models and makes are to be seen, with Fiat, Isuzu, Kia, LDV, Mercedes, Mitsubishi, Mudan and Toyota minibuses currently on the scene. The latest regulations now allow seating to be increased from 18 to 21 on the larger vehicles. However the majority of the operators still operate 14-seat minibuses.

Bristol LH DBY 303 displays a Leyland Tiger Cub badge as it heads through Floriana on a surprisingly quiet Sunday afternoon. Route 41 is a circular route from Valletta which serves Msida, Tal Qroqq, Kappara and Ta' Zwejt before reaching the terminus in San Gwann.

The white mini-buses are owned by rental companies and can be hired on a self-drive basis or can be used with a driver as transport for holiday companies, especially on airport-hotel transfers. All the white minibuses carry an xVY registration, but the letter combinations are varied and within these letter groups the number of vehicles can vary from one (OVY 002) to almost 150 within the LVY sequence. The following letter combinations are known to have been issued: AVY to FVY inclusive, HVY to LVY inclusive, OVY, RVY, SVY, UVY and WVY. Most combinations start with 001 and run for the most part consecutively, but there are variations. The latest regulations for these vehicles now allow seating for up to fourteen passengers from the previous ten.

Open-top buses have been allocated to new series: COY 001 to COY 012 are operated by Cancu Supreme Travel Ltd and Garden of Eden Garage Ltd operates COY 013 to 015. But, the latest Open-top buses carry the new xPY registration.

Minibuses for the disabled are equipped with tail lifts carry the xSY nnn registrations. Three different operators work in this sector, but all their vehicles have an overall light-blue livery.

Airport vehicles are rarely seen on public roads, and so few bear registrations, though two of the former Deutsche Bundesbahn ones did so, AMG 043 and AMG 044, in addition to their airport fleet number. The current airside bus fleet consists of at least eleven owned by Air Malta and three that belong to Globeground Malta.

Police buses carry GVP nnn registrations and the Education Department buses the GVH nnn and GVN nnn plates for mini-buses. The Army has a fleet of around 25 Iveco service buses with Cacciamali bodies that retain Italian Army number plates.

Furthermore there are several other NPSVs which include contract, hospitality, welfare and trainers and a number of stored double-deck buses.

ROUTE BUSES ON MALTA

The seven columns listing the details of each vehicle follow the same format throughout this *Bus Handbook*. The names of the "Registered Owners" have been obtained from offical sources and are those registered with the ATP, as at March 2009.

Accompanying these seven columns is a second section which presents a chronological list of the registrations held by the licence for each current plate. For all vehicles which have been imported from Great Britain, Eire or the Isle of Man, the registration(s) carried there are given after the Maltese series.

In the "Previous Owners" column, background information on the history of each vehicle is provided, wherever such details are known, including the original British or Irish owner of a vehicle. These details amplify any other information currently available in print.

Registration	Chassis	Bodywork	Seating	New	To Malta	Registered Owner
DBY 300	AEC Reliance 6MU2RA	Willowbrook	B46F	9/68	4/81	Charles Caruana, Mellieha
DBY 301	AEC Swift 4MP2RV	Marshall	B46F	1/71	12/80	Francis Galea, Fgura
DBY 302	AEC Reliance 2MU3RA	Ciantar (1978)	B45F	1960	4/78	Horace Vella, Qormi
DBY 303	Bristol LH6L	Eastern Coach Works	B45F	11/76	7/87	Dorothy Camilleri, Naxxar
DBY 304	Bristol LH6L	Eastern Coach Works	B45F	5/74	10/86	Noel Cassar, Zabbar
DBY 305	AEC Swift 4MP2R	Park Royal	B46F	10/70	2/82	Paul & Rocco (Marsa Ltd), Marsa
DBY 306	Leyland Royal Tiger PSU1/16	Aquilina (1973)	BC45F	5/53	4/73	C.M.S. Transport Ltd (Stephen Cilia), Gharghur
DBY 307	Dennis Dart SLF	Plaxton Pointer	N45F	5/97	5/97	Leo Grech, Mosta
DBY 308	AEC Mercury (Zammit)	Gauci	B40F	10/66	10/66	David Borg, Mellieha
DBY 309	Maltese Bedford f/c (Zammit)	Barbara	B40F	4/67	4/67	Saviour Vella, Zejtun
DBY 310	Bedford YRQ	Duple Dominant	BC45F	1/73	by 10/88	Charlie Abela, Zejtun
DBY 311	BMC Falcon	BMC	N45F	5/03	5/03	Mario Psaila, Zurrieq
DBY 312	King Long XMQ6113GMC	King Long	N45F	7/03	7/03	Raymond Buttigieg, Paola
DBY 313	Solaris Valletta	Solaris	N45F	5/03	5/03	John Ellul, Mqabba
DBY 314	Bedford YLQ	Duple Dominant I	BC45F	5/76	by 1/85	Carmelo Micallef, Qormi
DBY 315	Leyland Lynx	Leyland	N45F	11/89	2/08	C.M.S. Transport Ltd,
DBY 316	Bedford YLQ	Duple Dominant II	BC45F	1/78	3/86	Johan Tanti, Mqabba
DBY 317	King Long XMQ6113GMC	King Long	N45F	8/04	8/04	Francis Cassar, Marsascala
DBY 318	Maltese Bedford SL (Zammit)	Brincat (1972), Zinnu (1998/9)	B45F	6/72	6/72	Maria Rosaria Polidano, Kirkop
DBY 319	Bedford YRQ	Duple Dominant	BC45F	5/73	11/84	John Desira, Bir id-Deheb

Previous registrations in reverse of chronological order	Previous owner and/or original GB owner
DBY 300, Y-0300, Y-1001, A-0004, TNY495G	Cynon Valley 5
DBY 301, FBY 665, M-1530, EGN200J	London Transport SMS200
DBY 302, FBY 045, Y-0826, Y-0880, Y-1580, A-3117, 3117, YXD 11	Carmel Cauchi, Xewkija, Gozo: Edwards, Joys Green (This vehicle was in service on Gozo from 3/86 until 3/03)
DBY 303, Y-0303, OCA630P	Dragon Mechanical Services, Pembroke Dock: Crosville Motor Services, SLL630
DBY 304, Y-0304, PTT606M	Western National, 1606
DBY 305, Y-0305, Y-1006, A-0089, EGN 244J	London Transport, SMS244
DBY 306, Y-0306, Y-0879, Y-1579, A-2908, 2908, MUF650	Unscheduled bus, Malta (Y-0879): Southdown,1650
DBY 307	New
DBY 308, Y-0308, Y-1009, A-0117, 117	New, on an imported chassis
DBY 309, Y-0309, Y-1010, A-0140, 140	New, on an imported chassis
DBY 310, Y-0310, LBU153L	Newton, Guildford
DBY 311	New
DBY 312	New
DBY 313	New
DBY 314, Y-0314, LUX543P	Cyril Evans, Senghenydd: Corvedale, Ludlow, 43
DBY 315, G168EOG	Travel West Midlands, 1168
DBY 316, Y-0316, VNT48S	Parry, Pwllheli: Corvedale, Ludlow, 48
DBY 317	New
DBY 318, Y-0318, Y-1019, A-0236, 236	Maltese-built chassis
DBY 319, Y-0319, Y-1020, VBX518L	Glyn Williams, Lower Tumble, 52: Jenkins, Llanelli

DBY 332 is one of a small number of Volvo vehicles with locally-built Scarnif bodywork. "King of the Road" is on lay-over at Valletta Castle Gate bus terminus in September 2009.

DBY 320	Maltese (Zammit)	Debono	BC40F	by 12/68	10/70	Francis Attard, Luqa (Stephen Cilia), Gharghur
DBY 321	AEC Swift 4MP2R	Marshall	B53F	10/70	12/80	Mario Farrugia, Ghaxaq
DBY 322	Fordson	Tonna (1964)	B39F	1933	1933	Jason Bugeja, Zabbar
DBY 323	Bedford YRQ	Duple Dominant	C45F	4/73	by 4/85	Bartholomeo Muscat, Mgarr
DBY 324	Ford V8 lorry (f/c 1964)	Casha (1964)	B36F	1942	2/51	Stefan Sammut, San Gwann
DBY 325	Bedford SBO	Schembri (1954), Grech (1979)	B45F	12/52	6/54	Godfrey Mifsud, Mgarr
DBY 326	King Long XMQ6113GMC	King Long	N45F	1/05	1/05	Grezzju Borg, Zabbar
DBY 327	Maltese f/c	Aquilina	B36F	10/66	3/71	Sean Grech, Naxxar
DBY 328	Bedford YLQ	Plaxton Supreme III Express	BC45F	5/77	by 10/87	Vincent Camilleri, Mosta
DBY 329	Bedford OB f/c	Duple: Gauci (1964)	B40F	1/47	1/47	Justin Borg, Attard
DBY 330	Ford Cargo	Ramco (1999)	B45F	by 12/88	1/99	Alex Pulo, Paola
DBY 331	AEC Reliance MU3RA	Debono (1969)	BC45F	by 12/55	10/69	Z.S. Co. Ltd, Naxxar
DBY 332	Volvo B7RLE	Scarnif	N45F	3/04	3/04	Salvatore Vella, Zejtun
DBY 333	AEC Reliance	Plaxton Supreme IV (1979)	BC47F	2/68	4/86	Horace Vella, Qormi
DBY 334	Bedford YLQ	Plaxton Supreme III	BC45F	2/77	4/86	Charles Vella, Mosta
DBY 335	Bedford YRQ	Plaxton Elite Express II	BC45F	12/71	by 11/84	CMS Transport Ltd, Gharghur

DBY 320, Y-0320, Y-0887, Y-1587, A -0699, 699 — Unscheduled coach, Malta (Y-0887)
DBY 321, Y-0321, EGN152J — Ministry of Education, Malta: London Transport, SMS152 (joined route bus fleet in 5/94)
DBY 322, Y-0322, Y-1023, A-0360, 360 — New chassis
DBY 323, Y-0323, NLG105L — Sheffield Health Authority: Bullock of Cheadle
DBY 324, Y-0324, Y-1025, A-0454, 454 — Maltese lorry, 5170, previously Army truck, F6809
DBY 325, Y-0325, Y-1026, A-0529, 529 — New
DBY 326 — New
DBY 327, Y-0327, Y-0850, Y-0868, Y-1568, A-7209, 7209 — Unscheduled bus, Malta (Y-0850). Chassis assembled locally in 1966 but not bodied until 3/71.
DBY 328, Y-0328, SUJ974R — Glyn Evans, Manmoel: Hampson, Oswestry
DBY 329, Y-0329, Y-1030, A-0557, A-2059, 2059 — Route bus, Malta
DBY 330 — Rebuild on lorry chassis
DBY 331, Y-0331, Y-0878, Y-1578, A-5573, 5573, HWH450 — Unscheduled bus, Malta (Y-0878): Hargreaves, Bolton
DBY 332 — New
DBY 333, KCY884, Y-0884, HPG650V, EJP561F — Unscheduled bus, Malta (Y-0884): Thomas, West Ewell (rebuilt from AEC Reliance 6MU3R in 1978, rebodied 1979)
DBY 334, Y-0344, RUT390R — Nu-Venture, Maidstone
DBY 335, Y-0335, XAW326K — Williams, Cwmdu: Hampson, Oswestry

DBY 341 is a fine example of local Maltese craftsmanship. This Maltese f/c (Baileys) with Brincat bodywork is seen in the afternoon sun at the Golden Bay terminus of route 652 to Sliema.

DBY 336	Bedford QL	Gauci (1959)	B40F	1940s?	10/59	Samuel Camilleri, Mosta
DBY 337	Dennis Lance SLF 11m	Wright Pathfinder	B45F	4/94	11/07	Silvio Borg, Mellieha
DBY 338	Albion Chieftain CH51Y	Brincat (1976)	BC45F	9/67	7/76	Joseph Falzon, Mosta
DBY 339	Bedford YLQ	Duple Dominant II	C45F	8/79	by 4/88	Etienne Falzon, Siggiewi
DBY 340	King Long XMQ6113GMC	King Long	N45F	11/03	11/03	Charles Caruana, Mellieha
DBY 341	Maltese f/c (Baileys)	Brincat	B40F	10/63	11/64	Godfrey Mifsud, Mgarr
DBY 342	King Long XMQ6113GM	King Long	N45F	7/03	7/03	Joseph Polidano, Kirkop
DBY 343	Bedford YRQ	Duple Dominant I	BC45F	8/74	5/85	George Cutajar, Tarxien
DBY 344	Bedford YLQ	Duple Dominant II	C45F	5/78	c9/00	Filippa d'Amato, Zebbug
DBY 345	AEC Reliance 6MU2RA	Willowbrook	B45F	9/68	6/81	Antonio Borg, Attard
DBY 346	AEC Regal III 0962	Zammit (4/65)	B42F	1948	4/65	Etienne Vella, San Gwann
DBY 347	Albion Clydesdale	Brincat (1984)	BC45F	11/68	5/84	Francis Cortis, Naxxar
DBY 348	Dodge KT1050	Daina (1978)	BC45F	1969	10/81	Alfred Xerri, Qormi
DBY 349	AEC Regal 0662	Barbara (1960)	B40F	1939	3/60	Saviour Camilleri, Naxxar
DBY 350	AEC Mercury GM4RH	Farrugia	B40F	1956	5/65	Joseph Buhagiar, Zabbar

DBY 336, Y-0336, Y-1037, A-0777, 777, 7054 — Unscheduled bus, Malta (Y-1037): RAF chassis
DBY 337, L34WLH — UK North, Manchester: London Buses LLW34
DBY 338, Y-0338, Y-0854, Y-1554, A-7054, 7054 — Unscheduled bus, Malta (Y-0854); chassis originally built as a lorry for Elders and Fyffes Ltd, London in 7/67

DBY 339, Y-0339, JWU461V — Williams, Cwmdu: Steel, Addingham, 32
DBY 340 — New
DBY 341, Y-0341, Y-1042, A-0857, 857 — Built on a new "Impala"-type chassis (10/63)
DBY 342 — New
DBY 343, Y-0343, UJB726N — Reliance, Newbury, 142
DBY 344, PUO852S — Evans, Tregaron: Turner, Chumleigh
DBY 345, Y-0345, Y-1046, A-0915, TNY494G — Cynon Valley, 4
DBY 346, Y-0346, Y-1047, A-0921, 921, HYP309 — Catt and Swinn, Great Bromley
DBY 347, Y-0347, Y-0891, Y-1591 — Unscheduled bus, Malta (Y-0891): originally lorry chassis in UK imported in 1979 but not bodied until 1984
DBY 348, Y-0348, Y-0886, Y-1586, A-2128, 2128, SDD206G — Unscheduled bus, Malta (Y-0886): imported Dodge chassis and cab
DBY 349, Y-0349, Y-1050, A-0983, 983, HFC413 — City of Oxford, 6
DBY 350, Y-0350, Y-1051, A-0987, 987 — Zarb Coaches, Malta: AEC Mercury chassis, new in 1956 and acquired in 2/64

DBY 351	Bedford YRQ	Plaxton Panorama Elite II	BC45F	3/71	by 11/84	Nazzareno Caruana
						(AC Buses), Zabbar
DBY 352	Bristol LH6L	Eastern Coach Works	B45F	5/75	by 10/86	Rodianne Sultana, Santa Venera
DBY 353	Maltese f/c (Baileys)	Schembri	B40F	4/64	4/66	Stephen Scerri, Mosta
DBY 354	Bedford YLQ	Plaxton Supreme IV Express	BC45F	5/79	by 1/87	Horace Vella, Qormi
DBY 355	Ford R1014	Plaxton Supreme III	BC45F	2/78	by 4/85	Dennis Borg, Mqabba
DBY 356	King Long XMQ6113GMC	King Long	N45F	11/03	11/03	Saviour Camilleri, Naxxar
DBY 357	BMC Falcon	BMC	N45F	5/03	5/03	Paul Spiteri, Qrendi
DBY 358	Bedford YLQ	Plaxton Supreme III	BC45F	4/77	by 4/85	Jeffrey Camilleri, Qormi
DBY 359	Bedford YLQ	Duple Dominant II	BC45F	2/78	by 4/85	Tal-Linja Co. Ltd, Naxxar
DBY 360	Bristol LH6L	Eastern Coach Works	B45F	12/73	by 11/95	Silvio Camilleri, Naxxar
DBY 361	Bedford QLD	Sammut	B40F	1944	3/58	Ronald Demanuele, Zabbar
DBY 362	Bedford YRQ	Duple Dominant	BC45F	12/74	by 10/86	Jonathan Mallia, Lija
DBY 363	King Long XMQ6113GM (m)	King Long	N45F	11/03	11/03	George Cutajar, Tarxien
DBY 364	Volvo B6-50	Alexander Dash	B44FL	1/95	2/08	Emanuel Cutajar, Gharghur
DBY 365	Maltese f/c (Baileys)	Schembri (6/64)	B40F	6/64	6/64	Horace Vella, Qormi
DBY 366	King Long XMQ6113GMC	King Long	N45F	11/03	11/03	Anton Gatt, Zejtun
DBY 367	Dennis Dart SLF	East Lancs Spryte	N45F	3/97	3/97	Leo Grech, Mosta

DBY 351, Y-0351, XVY198J Tozer, Scarborough, 4: York Pullman, York
DBY 352, Y-0352, HBX948N Davies Bros., Pencader, 97
DBY 353, Y-0353, Y-1054, A-1058, 1058 New chassis supplied to Gauci of Qormi
DBY 354, Y-0354, TEJ102T Evans, Penrhyncoch
DBY 355, Y-0355, XEF936S, 3996 EL, XHS122S Drew, Guisborough: Wilson, Carnwath
DBY 356 New
DBY 357 New
DBY 358, Y-0358, UWU628R Collin, East Hardwick
DBY 359, Y-0359, BTU655S Bostock, Congleton, 10
DBY 360, Y-0360, Y-0705, NFJ596M Western National, 1576
DBY 361, Y-0361, Y-1062, A-1181, 1181 4x4 dropside military lorry chassis
DBY 362, Y-0362, STM725N Ouse Valley, Goldington
DBY 363 New
DBY 364, M512TRA Stagecoach 30230: Nottingham Transport, 512
DBY 365, Y-0365, Y-1066, A-1290, 1290 Route bus, Malta
DBY 366 New
DBY 367 New

One of the more recent vehicles to enter service in the route bus fleet on Malta is DBY 364, a Volvo B6 with Alexander Dash bodywork, which was formerly with Stagecoach. It awaits its next departure from Qawra bus terminus on the main route 49 to Valletta during the late afternoon in September 2009.

BMC Falcon DBY 370 is parked on the bus lay-bys in front of the main terminal buildings at Malta International Airport at Luqa. It was rostered to operate the Gozo to Airport Shuttle service and was waiting for its next departure at 1410 in August 2008.

DBY 368	Reo Speedwagon	Aquilina; (Sammut 1955)	B36F	5/39	5/39	Mario Farrugia, Attard
DBY 369	AEC Reliance 2MU3RA	Spiteri (1983)	B46F	1961	1/83	Peter Pace, Qrendi
DBY 370	BMC Falcon	BMC	N45F	5/03	5/03	Paul Mifsud, Mellieha
DBY 371	Bedford SB8	Debono	B40F	4/62	4/62	Francis Galea, Fgura
DBY 372	Ford ET7 f/c	Gasan (BMC Garage); (Casha 1970)	B40F	3/55	3/55	Maria Antonia Zammit, Qormi
DBY 373	Bedford YRQ	Scarnif (1994)	B45F	9/74	by 10/86	Mary Scicluna, Ghaxaq
DBY 374	AEC Mercury 2GM4RH	Barbara	B40F	1958	3/68	Grezzju Borg, Zabbar
DBY 375	Bedford SBG	Aquilina (1966)	B41F	by 12/55	4/66	Ramon Caruana, Fgura
DBY 376	Bedford SB	Mifsud	B40F	by 12/69	by 12/69	Anthony Falzon, Qormi
DBY 377	King Long XMQ 6113GMC	King Long	N45F	6/03	6/03	Lorenza Buttigieg, Zabbar
DBY 378	Leyland Tiger Cub	East Lancs	B46F	1966	5/81	Horace Vella, Qormi
DBY 379	Leyland Lynx	Leyland	BC45F	3/89	2/08	Saviour Grech, Mqabba
DBY 380	AEC Swift 4MP2R1	Park Royal	B46F	9/70	7/81	Paul & Rocco (Marsa Ltd), Marsa
DBY 381	AEC Mercury	Brincat	B40F	7/67	7/67	Paul Cardona, Mellieha
DBY 382	King Long XMQ 6113GM (m)	King Long	N45F	5/03	5/03	John Borg, Mellieha
DBY 383	Bedford YRQ	Plaxton Elite II	BC45F	10/71	by 4/85	Winstin Muscat, St Paul's Bay
DBY 384	Maltese f/c	Barbara	B40F	11/68	11/68	Charlie Borg, Mellieha
DBY 385	Bedford YLQ	Plaxton Supreme III Express	C45F	4/77	by 1/87	Charlie Abela, Zejtun
DBY 386	Bedford SB1	Schembri (5/61); Sladden (2/77)	B40F	5/61	5/61	Angelo Abela, Dingli
DBY 387	This registration is currently not in use					Country Boy Ltd, Birkirkara
DBY 388	King Long XMQ113GM	King Long	N45F	10/03	10/03	Mario Jasmine Farrugia, Tarxien
DBY 389	AEC Swift 4MP2R1	Park Royal	B47F	1/71	7/81	Alexander Borg, Rabat
DBY 390	King Long XMQ 6113GMC	King Long	N45F	11/03	11/03	Carmel Abela, Zejtun
DBY 391	Bedford SBO	Debono	B36F	2/56	2/56	Andrew Abela, Zejtun
DBY 392	Bedford SBG	Aquilina(?) (1957): Barbara (1968)	B40F	11/57	11/57	Donald Vince Borg, Mosta
DBY 393	King Long XMQ 6113	King Long	N45F	4/03	4/03	Paul Cutajar, Siggiewi
DBY 394	BMC Falcon	BMC	N45F	5/03	5/03	Raymong Buttigieg, Zabbar
DBY 395	Maltese (Baileys)	Casha	B46F	8/62	8/62	Grezzju Borg, Zabbar
DBY 396	Bedford QLR	Debono (5/59)	B36F	1943	5/59	Paul Cassar, Zabbar
DBY 397	GMC f/c	Tonna (1948): Xuereb (1954)	B40F	1943	10/48	Siliano Sammut, Mosta
DBY 398	Leyland Lynx	Leyland	B45F	3/89	2/08	Francis Galea, Fgura

DBY 392 is a Bedford SBG with Barbara bodywork constructed in 1968 but has been in service for well over fifty years. It is seen at Qawra bus terminus in an evening waiting to depart to Golden Bay on route 51. This unusual short-working operates, often only on demand, in the early evening to relieve over-crowding on the 652 service to Sliema.

DBY 368, Y-0368, Y-1069, A-1306, 1306, 2370	Route bus, Malta
DBY 369, Y-0369, Y-0903, Y-1603, A-0434, 434, RDB860	Zarb Coaches, Malta (Y-0903): North Western, 860
DBY 370	New
DBY 371, Y-0371, Y-1072, A-1390, 1390	New
DBY 372, Y-0372, Y-1073, A-1403, 1403	n/c route bus on St Julian's - Birkirkara route, Malta
DBY 373, Y-0373, SYO601N	Wilder, Feltham
DBY 374, Y-0374, Y-1075, A-1418, 1418	Galea, Fgura (certified as late as 3/68)
DBY 375, Y-0375, Y-1076, A-1421, 1421	Royal Navy, 33RN78
DBY 376, Y-0376, Y-0869, Y-1569, A-0232, 232	Unscheduled bus, Malta (Y-0869)
DBY 377	New
DBY 378, Y-0378, Y-1079, A-1526, YTD290D	Blackburn, 10
DBY 379, F213YHG	Preston Bus, 213
DBY 380, Y-0380, Y-1081, A-1585, EGN232J	London Transport, SMS232
DBY 381, Y-0381, Y-1082, A-1604, 1604	New as an AEC Mercury chassis assembled in Malta and bodied as a bus.
DBY 382	New
DBY 383, Y-0383, Y-1084, AJH128K	Bantam, Ipswich: North Star, Stevenage
DBY 384, Y-0384, Y-1085, A-1643, 1643	New
DBY 385, Y-0385, RBO669R	Williams, Cwmdu: Hills of Tredegar
DBY 386, Y-0386, Y-0562, Y-0853, Y-1553, A-2681, 2681	Unscheduled bus, Malta on new Bedford SBI chassis
DBY 387	
DBY 388	New
DBY 389, Y-0389, Y-1090, A-1751, EGN329J	London Transport, SMS329
DBY 390	New
DBY 392, Y-0392, Y-1093, A-1764, 1764	Imported Bedford SBG chassis in 6/57
DBY 393	New
DBY 394	New
DBY 395, Y-0395, Y-1096, A-1784, 1784	New chassis
DBY 396, Y-0396, Y-1097, A-1793, 1793	Army Bedford QLR 4x4 radio lorry
DBY 397, Y-0397, Y-1098, A-1795, 1795	1943 GMC military lorry chassis
DBY 398, F211YHG	Preston Bus, 211

An early morning trip finds MAN Scarnif DBY 399 on lay-over in Qawra bus terminus. Route 580 is one of the comparatively new services to the Mater Dei Hospital and University. About one dozen route buses now carry locally built Scarnif bodywork using a variety of chassis types.

DBY 399	MAN 14.220	Scarnif	N45F	6/03	6/03	Joseph Mallia, Lija
DBY 400	Bedford YRQ	Plaxton Panorama Elite Exp III	BC45F	3/73	8/85	Pasquale Sciberras, Mqabba
DBY 401	Leyland TS4	Leyland (1932); Debono (1965)	B40F	1932	8/48	Francis Cutajar, Zabbar
DBY 402	Leyland Royal Tiger PSU1/15	Harrington (1952): Aquilina (1974)	B44F	10/52	8/73	Etienne Falzon, Siggiewi
DBY 403	BMC Falcon	BMC	N45F	5/03	5/03	John Zammit, Qormi
DBY 404	Bedford SBO	Sammut	B40F	1954	by 11/95	Francis Attard, Luqa
DBY 405	Ford R192	Plaxton Panorama Elite III	BC46F	3/74	by 9/85	Francis Galea, Fgura
DBY 406	Bedford YRQ	Plaxton Panorama Elite Exp III	BC45F	8/75	by 6/89	Paul & Rocco (Marsa Ltd), Marsa
DBY 407	King Long XMQ 6113GMC	King Long	N45F	10/03	10/03	Tarcisio Gatt, Marsascala
DBY 408	AEC Mercury	Farrugia	B40F	9/63	1/68	Joseph Refalo, Hamrun
DBY 409	Bedford SB1	Debono	B40F	12/59	12/59	Raymond Borg, Mellieha
DBY 410	Bedford YLQ	Duple Dominant II	C45F	4/78	by 4/89	Joseph Farrugia, San Gwann
DBY 411	Bedford YRQ	Duple Dominant I	C45F	4/74	by 4/85	Paul & Rocco (Marsa Ltd), Marsa
DBY 412	Bedford YRQ	Duple Dominant I	C45F	5/74	by 1/87	Samuel Muscat, Mgarr
DBY 413	BMC Falcon	BMC	N45F	10/03	10/03	Carmel Borg, Mellieha
DBY 414	Bedford YLQ	Duple Dominant I	C45F	3/77	11/84	Michael Micallef, Qormi

DBY 391, Y-0391, Y-1092, A-1258, 1258 — Route bus, Malta
DBY 399 — New
DBY 400, Y-0400, CPY580L — Carnell, Sutton Bridge: Walker, Brompton
DBY 401, Y-0401, Y-1102, A-1823, 1823, 3356 — Possibly a DD from NE England
DBY 402, Y-0402, Y-0885, Y-1585, A-2846, 2846, OTE915 — Zarb Coaches, Malta (Y-0885): Gregson, Skelmersdale
DBY 403 — New
DBY 404, Y-0404, Y-0557, Y-0595, Y-1296, A-2765, 2765 — Route bus, Malta
DBY 405, Y-0405, VYC869M — Andrews, Trudoxhill
DBY 406, Y-0406, KUR304P — Fletcher, Skelmersdale: Grogan, Rainhill
DBY 407 — New
DBY 408, FBY 769, Y-0769, Y-1470, A-3346, 3346 — Route bus, Malta; AEC Mercury chassis imported in 9/63
(This vehicle re-entered service in 2004, following the destruction by fire of the previous Bedford SB, DBY 408)
DBY 409, Y-0409, Y-1110, A-1861, 1861 — New
DBY 410, Y-0410, TPJ274S — Ron's, Ashington: Safeguard, Guildford
DBY 411, Y-0411, Y-1112, XTH700M — Ffoshelig Motors, Newchurch
DBY 412, Y-0412, VNK480M — Reid, Bedford: Mason, Perivale
DBY 413 — New
DBY 414, Y-0414, Y-1115, STT962R — Cyril Evans, Senghenydd: Kingdom, Tiverton

King Long DBY 421 is seen at Valletta's Castle Gate terminus, as it wends its way round to the departure stand for route 145 to Cirkewwa one lunch-time.

DBY 415	King Long XMQ 6113GMC	King Long	N45F	11/03	11/03	Albert Camilleri, Naxxar
DBY 416	King Long XMQ 6113GM (m)	King Long	N45F	8/03	8/03	Filippa d'Amato, Zebbug
DBY 417	Bedford YRQ	Plaxton Elite III	BC45F	4/75	9/84	Gaetano Sciberras, Kirkop
DBY 418	Bedford YLQ	Plaxton Supreme III	C45F	1/77	by 9/85	Albert Sammut, San Gwann
DBY 419	Leyland Hippo	Micallef (1948); Scarnif (1995)	B45F latest 12/40		2/48	John Mary Cachia, Luqa
DBY 420	AEC Swift 4MP2R1	Park Royal	B46F	4/71	7/81	Silvio Buttigieg, Fgura
DBY 421	King Long XMQ 6113GMC	King Long	N45F	10/03	10/03	Gerald Camilleri, Naxxar
DBY 422	Bristol LH6L	Eastern Coach Works	B45F	2/74	by 3/86	Josephine Cortis, Naxxar
DBY 423	King Long XMQ 6113GM (m)	King Long	N45F	2/04	2/04	Noel Vella, Mellieha
DBY 424	Bedford YRQ	Duple Dominant Express	BC45F	6/74	by 9/85	Albert Sammut, San Gwann
DBY 425	Bedford YLQ	Duple Dominant II	BC45F	2/78	by 4/85	Alexander Bonavia, Naxxar
DBY 426	AEC Swift 4MP2R	Marshall	B46F	6/70	6/81	Rosaria Falzon, Fgura
DBY 427	Leyland Leopard PSU4B/2R	Marshall	B46F	11/72	10/90	Paul & Rocco (Marsa Ltd), Marsa
DBY 428	Dodge f/c	Casha	B40F	1956	1956	Albert Sammut, San Gwann
DBY 429	MAN 14.220	Scarnif	N45F	7/03	7/03	Patrick Cauchi, Mellieha
DBY 430	King Long XMQ 6113GMC	King Long	N45F	7/03	7/03	Mary Buttigieg, Floriana

DBY 415	New
DBY 416	New
DBY 417, Y-0417, Y-1118, KCK539N	Parfitt's, Rhymney Bridge: Grange, Morecambe
DBY 418, Y-0418, RRR905R	Mowbray, Stanley: Redfern, Mansfield
DBY 419, Y-0419,Y-0321, Y-0646, Y-1347, A-2981, 2981	War Department lorry, fleet number N55. It also carried Y-0466 and Y-0559 for short periods
DBY 420, Y-0420, Y-1121, A-1952, EGN427J	Leybourne Grange Hospital, West Malling: London Transport, SMS427
DBY 421	New
DBY 422, Y-0422, NLJ523M	Hants and Dorset, 3523
DBY 423	New
DBY 424, Y-0424, SNX601M	Tanat Valley, Pentrefelin: Monty Moreton, Nuneaton
DBY 425, Y-0425, BTU654S	Bostock, Congleton, 33
DBY 426, Y-0426, Y-1127, A-1971, AML15H	London Transport, SM15
DBY 427, Y-0427, GKE457L	Wealden, Five Oak Green: Maidstone and District, 3457
DBY 428, Y-0428, Y-0480, Y-1181, A-2237, 2237	Route bus, Malta
DBY 429	New
DBY 430	New

With the Phoenicia Hotel on the left and the Triton Fountain on the right, DBY 445, an Albion Victor with Barbara bodywork, leaves Valletta's Castle Gate bus terminus on route 81 to Dingli in March 2009.

DBY 431	Bedford YRQ	Duple Dominant Bus	B47F	3/76	by 6/91	Massimo Bonella, Bugibba
DBY 432	King Long XMQ 6113GMC	King Long	N45F	10/03	10/03	Francis Galea, Fgura
DBY 433	Solaris Valletta	Solaris	N45F	12/02	12/02	Mario Mifsud, Mgarr
DBY 434	Indiana f/c	Aquilina (8/54): rebuilt 2007	B40F	5/38	5/38	Adrian Buttigieg, Zabbar
DBY 435	AEC Mercury (Zammit)	Barbara	B40F	3/66	3/66	John Abela, Zejtun
DBY 436	Bedford SB	Casha	B40F	1952	by 8/74	Carmel Cassar, Kalkara
DBY 437	Ford R1014	Plaxton Supreme III	C45F	8/78	by 1/87	John Desira, Zabbar
DBY 438	AEC Mercury (Zammit)	Farrugia	B40F	2/67	2/67	Anthony Falzon, Qormi
DBY 439	This vehicle was burnt out in 12/08					Carmelo Vella, Birzebbuggia
DBY 440	MCII	MCII	N45F	4/03	4/03	Joe Abela, Birzebbugia
DBY 441	Bedford YRQ	Plaxton Elite III	BC45F	6/74	by 1/85	John Abela, Zejtun
DBY 442	Bedford YLQ	Plaxton Supreme III	BC45F	1/77	by 8/86	John Xerri, Mosta
DBY 443	King Long XMQ 6113GM (m)	King Long	N45F	10/03	10/03	Alex Farrugia, Mgarr
DBY 444	Leyland Royal Tiger PSU1/13	East Lancs/Scarnif (2000)	B45F	2/67	by 6/82	Joseph Buhagiar, Zebbug
DBY 445	Albion Victor VT17AL	Barbara	B40F	9/60	8/67	Michael Pace, Naxxar

DBY 431, Y-0431, NKE303P	Penjon, Benenden:
	Maidstone Borough Council Transport Department, 3
DBY 432	New
DBY 433, EBY 433	New
(EBY 433 was issued incorrectly when this vehicle was first registered; DBY 433 was issued in 10/05)	
DBY 434, Y-0434, Y-1135, A-2002, 2002	n/c route bus, Malta on chassis of 1937 vintage
DBY 435, Y-0435, Y-1136, A-2003, 2003	Chassis imported from London and assembled locally
DBY 436, Y-0436, Y-1137, A-2007, 2007	Ministry of Supply
DBY 437, Y-0437, SUO247T	Evans, Tregaron: Lovering, Combe Martin
DBY 438, Y-0438, Y-1139, A-2018, 2018	Chassis imported from London and assembled locally
DBY 439	
DBY 440	New
DBY 441, Y-0441, Y-1142, SHE199M	Morley, West Row: Mawby, Crowcroft and Fell, Barnsley
DBY 442, Y-0442, RGS98R	McAndrew, Leamington Spa: Stanley, Hersham
DBY 443	New
DBY 444, Y-0444, Y-1145, GBV17E	Blackburn, 17
DBY 445, Y-0445, Y-1146, A-2083, 2083	New (stored unbodied between 1960 and 1967)

With the Hilton Hotel behind, King Long DBY 448 passes the Paceville terminus heading for Bahar ic-Caghaq on route 68 in the early afternoon of 7th September 2009.

DBY 446	King Long XMQ 6113GMC	King Long	N45F	11/03	11/03	KristinoGiordmaina, Siggiewi
DBY 447	Bedford YRQ	Duple Dominant I	BC45F	8/75	by 8/86	John Mary Sacco, Qrendi
DBY 448	King Long XMQ 6113GMC	King Long	N45F	6/03	6/03	Adrian Buttigieg, Zabbar
DBY 449	Bedford YRQ	Duple Dominant Express	BC45F	9/75	3/86	Joseph Abela, Tarxien
DBY 450	AEC Mercury	Aquilina	B40F	1958	2/66	Daniel Vella, Mosta
DBY 451	Bedford YMQ	Plaxton Derwent II	B45F	6/87	2/07	Saviour Abela, Birzebbugia (Swallow Garages Ltd)
DBY 452	King Long XMQ 6113GMC (m)	King Long	N45F	6/03	6/03	Anthony Buttigieg, Paola
DBY 453	Bedford YRQ	Plaxton Elite III	DP45F	1/73	by 10/86	Paul & Rocco (Marsa Ltd), Marsa
DBY 454	Volvo B7RLE	Saracakis	N45F	9/03	9/03	Victor Mifsud, Mgarr
DBY 455	Bedford f/c / Maltese Docks	Farrugia	B40F	5/70	1/72	Raymond Buttigieg, Zabbar
DBY 456	BMC Falcon	BMC	N45F	8/03	8/03	Carmel Cassar, Kalkara
DBY 457	Bedford YLQ	Plaxton Supreme III	BC45FL	1/77	by 8/86	Nicholas Grech, Mqabba
DBY 458	Leyland Royal Tiger PSU1/15	Aquilina (9/72)	B44F	1951	9/72	Francis Galea, Fgura
DBY 459	Ford R1014	Plaxton Supreme IV	BC45F	6/80	by 4/89	Emmanuele d'Amato, Siggiewi
DBY 460	Bedford SB	Brincat	B41F	7/54	7/54	Josef Grech, Gzira
DBY 461	King Long XMQ 6113GMC (m)	King Long	N45F	11/03	11/03	Rodianne Sultana, Santa Venera

DBY 446	New
DBY 447, Y-0447, JVJ442P	Alpha, Brighton
DBY 448	New
DBY 449, Y-0449, JWO48P	Bebb, Llantwit Fardre, 48
DBY 450, Y-0450, Y-0894, Y-1594, A-2625, 2625	Unscheduled bus, Malta (Y-0894) Chassis imported second-hand from Britain in 7/64
DBY 451, D136XVW	Hylton and Dawson, Leicester: Hedingham and District, L136
DBY 452	New
DBY 453, Y-0453, FNL581L	Waldron and Grange, Nottingham: Moordale, Newcastle
DBY 454	New
DBY 455, Y-0455, Y-0890, Y-1590, A-0634, 634	Unscheduled bus, Malta (Y-0890): imported chassis
DBY 456	New
DBY 457, Y-0457, SEH938R	Ganal, Shotts: Turner, Brown Edge
DBY 458, Y-0458, Y-0898, Y-1598, A-3326, 3326, DRN703	Zarb Coaches, Malta:Taylor Woodrow: Ribble, 781
DBY 459, Y-0459, GYJ955V	Dunn-Line, Nottingham: Terminus, London W1
DBY 460, Y-0460, Y-0910, Y-1610, A-4873, 4873	Unscheduled bus, Malta (Y-0910)
DBY 461	New

EBY 482 is a Dodge T100L with Aquilina bodywork, which has been in service for over fifty-four years. It is seen passing through the narrow section of main road in St Paul's Bay on the approach to the roundabout at Xemxija. Route 45 is the lengthy route from Valletta to Cirkewwa, the departure point for the ferry to Gozo.

DBY 462	Dennis Dart SLF 10.6m	Plaxton Pointer	N45F	5/97	5/97	Leo Grech, Mosta
DBY 463	Bedford SBO	Barbara	B46F	1955	1955	Horace Vella, Qormi
DBY 464	King Long XMQ 6113GMC (m)	King Long	N45F	6/03	6/03	Raymond Buttigieg, Zabbar
DBY 465	Bedford SB	Borg (9/66)				
		Sladden (1974/75)	B41F	4/55	9/66	Carmel Grech, Zabbar
DBY 466	AEC Swift 4MP2R	Marshall	B46F	4/70	9/83	Carmela Abela, Ghaxaq
DBY 467	Bedford YRQ	Marshall	B46F	7/75	by 3/90	Mario Abela, Zejtun
DBY 468	Bedford SB	Brincat (1972)	B40F	1956	1972	Paul & Rocco (Marsa Ltd), Marsa
DBY 469	Bristol LH6L	Eastern Coach Works	B45F	12/74	by 9/85	Maria Antonia Zammit, Qormi
DBY 470	Bristol LH6L	Eastern Coach Works	B45F	12/74	by 8/86	Marco Borg, San Gwann
EBY 471	Bedford f/c (Zammit)	Aquilina (1969)	B46F	10/65	2/69	Manuel Cassar, Zabbar
EBY 472	Dodge T110L f/c	Casha	B45F	1942	12/55	Alfred Mercieca, Marsa
EBY 473	Ford R1014	Caruana (2001)	B45F	7/78	by 6/92	Country Boy Ltd, Birkirkara
EBY 474	Bedford YRQ	Duple Dominant I	BC45F	1/75	by 11/84	Mario Buhagiar, Zabbar
EBY 475	Bedford YRQ	Plaxton Supreme III	C45F	4/76	by 1/87	Saviour Zahra, Rabat
EBY 476	Bristol LH6L	Eastern Coach Works	B45F	1/75	by 3/90	Carmel Abela, Zejtun
EBY 477	Bedford YLQ	Plaxton Supreme III	BC45F	8/78	by 1/87	Carmel Xerri, Qormi
EBY 478	This vehicle was burnt out in 8/08					Nicholas Fenech, Siggiewi
EBY 479	Leyland Leopard PSU4/4R	Marshall	B45F	4/68	by 5/81	Michael Cutajar, Kirkop
EBY 480	Bedford YLQ	Duple Dominant I	BC45F	8/76	by 10/87	Carmel Borg, Rabat
EBY 481	Bedford SBG	Mercieca (1978)	B45F	9/56	6/78	Paul & Rocco (Marsa Ltd), Marsa
EBY 482	Dodge T110L f/c	Aquilina (1955)	B40F	1942	12/55	Noel Vella, Mellieha
EBY 483	MAN 14.220	Scarnif	N45F	2/04	2/04	Godfrey Mifsud, Mgarr
EBY 484	AEC Regal	Debono (1963)	B40F	1948	5/63	Carmelo Buttigieg, Zabbar
EBY 485	Bedford YLQ	Plaxton Supreme III	BC45F	9/78	by 10/87	Horace Vella, Qormi
EBY 486	Ford V8 f/c	Borg (1948): Vella (1959)	B36F	1944	5/48	Reuben Galea, Msida
EBY 487	King Long XMQ 6113GMC	King Long	N45F	7/03	7/03	Victor Camilleri, Paola
EBY 488	MAN 14.220 HOCL-NR	Scarnif	N45F	12/02	12/02	Francis Attard, Luqa
EBY 489	King Long XMQ 6113GMC	King Long	N45F	11/03	11/03	Rodianne Sultana, Santa Venera
EBY 490	Ford R1014	Plaxton Supreme IV	C45F	1/79	by 7/88	Joseph Abela, Birzebbugia
						(Peppin Garage Ltd)
EBY 491	Bedford YRQ	Plaxton Elite Express III	BC45F	7/75	by 4/85	Mario Scerri, Siggiewi
EBY 492	Albion Chieftain CH13	Aquilina (1975)	B41F	12/62.	1/75	Silvio Buttigieg, Fgura
EBY 493	Bedford YRQ	Duple Dominant I	BC45F	8/74	by 4/85	Francis Galea, Fgura

Malta and Gozo Buses

EBY 479 is Leyland Leopard with a Marshall body which started life with West Yorkshire PTE . Its distinctive front with a predominance of white paintwork provides a marked contrast to many other route buses. It is parked at the boarding point of route 3 to Senglea in Castle Gate bus terminus in August 2008.

DBY 462	New
DBY 463, Y-0463, Y-0459, Y-1160, A-2135, 2135	Route bus, Malta (Y-0459)
DBY 464	New
DBY 465, Y-0465, Y-1166, A-2164, 2164	Ministry of Supply, fleet number 199, then NGY740 Strachan B55D 3+2 wooden slat seating, 4/55
DBY 466, Y-0466, M-1589, AML18H	Kalafrana Construction Co, Malta;: London Transport, SM18 imported 4/81 to Route Bus fleet 1/91
DBY 467, Y-0467, HGM613N	Atomic Weapons Research Establishment, Aldermaston
DBY 468, Y-0468, Y-0908, Y-1608, A-2139, 2139	Unscheduled bus, Malta (Y-0908) on chassis new in 1956
DBY 469, Y-0469, GDV464N	Southern National, 107
DBY 470, Y-0470, GDV458N	Western National, 1609
EBY 471, Y-0471, Y-1172, A-2217, 2217	Route bus, Malta on an locally built Bedford chassis
EBY 472, Y-0472, Y-1173, A-2218, 2218	Route bus by 1955, on Dodge truck chassis, 5175
EBY 473, XBH859S	Town and Around, Folkestone: White, Colnbrook
EBY 474, Y-0474, GUX402N	Squirrel, Hitcham: Corvedale, Ludlow
EBY 475, Y-0475, A111MAN, SBV284P	Tours Isle of Man, Douglas, 20: Florence, Morecambe
EBY 476, Y-0476, GLJ487N	I. M. Museum, Southsea: Hants and Dorset, 3555
EBY 477, Y-0477, YRY509T	Evans, Tregaron: Lester, Long Whatton
EBY 478	
EBY 479, Y-0479, Y-1180, A-2236, NHE10F	West Yorkshire PTE, 3360
EBY 480, Y-0480, Y-1181, OWV439R	Plumpton Coaches: Alpha, Brighton
EBY 481, Y-0481, Y-1182, A-2238, 2238	Royal Navy, 39RN97 in 1966, bodied in 1978
EBY 482, Y-0482, Y-1183, A-2247, 2247	Dodge truck, 5541
EBY 483	New
EBY 484, DBY 312, Y-0312, Y-1013, A-0158, 158, GNY764	Route bus, Malta: Thomas, Port Talbot
This vehicle re-entered service in 2/06 after the previous vehicle carrying this registration was scrapped.	
EBY 485, Y-0485, ANR741T	Fourways, Leeds
EBY 486, Y-0486, Y-0858, Y-1558, A-0192, 192	Unscheduled bus, Malta (Y-0858): War Department truck; F1068
EBY 487	New
EBY 488	New
EBY 488	New
EBY 489	New
EBY 490, Y-0490, DRW908T	Tedd, Winterslow: Smith, Wilnecote
EBY 491, Y-0491, LBR270N	Killay, Swansea: Gypsy Queen, Langley Park
EBY 492, Y-0492, Y-0864, Y-1564, A-2829, 2829	Welcome Garage, Malta (Y-0864):
Chassis originally ordered by Warwick Motors, Stoke-on-Trent on 19/9/62	
EBY 493, Y-0493, UJB725N	Reliance, Newbury, 141

One of the many British coaches exported to the islands during the 1980s is EBY 504, a Ford R1014 with Duple Dominant III bodywork. It is heading back to Sliema Ferries from Cirkewwa on route 645 and is passing the Paceville terminus of route 62.

EBY 494	BMC Falcon	BMC	N45F	5/03	5/03	Mario Ellul, Mqabba
EBY 495	Bedford YMQ	Marshall	B49F	12/81	by 10/91	Alphonse Abela, Ghaxaq
						(Peppin Garage Ltd)
EBY 496	Bedford YRQ	Plaxton Panorama Elite III	B45F	3/75	by 8/86	John Mary Ciappara, Birkirkara
EBY 497	Bedford YRQ	Plaxton Panorama Elite III	B45F	5/75	by 4/85	Justin Borg, Attard
EBY 498	AEC Reliance MU3RV	Aquilina (1973)	BC41F	5/56	10/73	Christopher Caruana, Zabbar
EBY 499	Ford R1015	Duple Dominant Bus	B45F	9/83	4/01	Charles Falzon, Mosta
EBY 500	Ford V8	Wayne (3/48);				
		Mercieca (1976)	B40F	9/47	3/48	Joseph Grech, Luqa
EBY 501	BMC Falcon	BMC	N45F	5/03	5/03	Tal-Linja Co. Ltd, Naxxar
EBY 502	Bedford SBG	Brincat	B39F	10/56	7/70	Frank Scicluna, Zabbar
EBY 503	Bedford YRQ	Duple Dominant	C45F	8/75	by11/84	Alphonse Abela, Ghaxaq
						(Peppin Garage Ltd)
EBY 504	Ford R1114	Duple Dominant II	B45F	5/78	3/86	Francis Cassar, Marsascala
EBY 505	Bedford YRQ	Plaxton Supreme III Express	C45F	3/76	by 1/87	Saviour Grech, Mqabba
EBY 506	Ford V8 f/c	Casha	B39F	1941	by 12/49	Jason Calleja, Mellieha
EBY 507	King Long XMQ 6113GMC	King Long	N45F	10/03	10/03	Albert Sammut, San Gwann
EBY 508	Bedford YLQ	Duple Dominant II	B45F	3/78	by 6/89	Francis Galea, Fgura

EBY 494
EBY 495, Y-0495, WVG787X
EBY 496, Y-0496, HKP219N
EBY 497, Y-0497, JWX93N
EBY 498, Y-0498, Y-0902, Y-1602, A-2673, 2673, VWE258
EBY 499, A666 BDL, A994 ADL
EBY 500, Y-0500, Y-1201, A-2329, 2329, 1058

EBY 501
EBY 502, Y-0502, Y-1203, A-2337, 2337

EBY 503, Y-0503, KYS415P
EBY 504, Y-0504, Y-0913, YNX447S
EBY 505, Y-0505, LRG66P
EBY 506, Y-0506, Y-1207, A-2357, 2357
EBY 507
EBY 508, Y-0508, GHG115S

New
Norfolk County Council Education Department
West Kent, Biggin Hill
Lockey, Bishop Auckland: Wallace Arnold, Leeds
Unscheduled bus, Malta (Y-0902): Sheffield United Tours
Isle of Wight County Council, Carisbrooke
Route bus 1058, with Wayne body; withdrawn 1966
and stored unbodied until 1976
New
Near East Air Force Command: AHQ Malta:
RAF, 32AC75 with Mulliner B31F body:
Burnside, Church Warsop: Nimmo, West Kilbride
Unscheduled bus, Malta (Y-0913): Central, Walsall
Williams, Cwmdu: Rochester & Marshall, Great Whittington
Route bus, Malta
New
Wilson, Failsworth: Holmes, Garstang

The frontal appearance of King Long EBY 511 is in complete contrast to that of DBY 448 on page 27. It awaits its next return journey to Valletta on route 67 and is seen at the St Andrew's terminus at about midday.

EBY 509	King Long XMQ 6113GM (m)	King Long	N45F	9/04	9/04	Raymond Cassar, Paola
EBY 510	King Long XMQ 6113GMC	King Long	N45F	10/03	10/03	Carmel Mifsud, Mgarr
EBY 511	King Long XMQ 6113GMC	King Long	N45F	5/03	5/03	Eugenio Zammit, Qormi
EBY 512	King Long XMQ 6113GMC	King Long	N45F	12/03	12/03	Francis Galea, Fgura
EBY 513	Bedford YRQ	Plaxton Elite II	B45F	5/72	3/85	Mario Zarb, Birkirkara
EBY 514	King Long XMQ 6113GMC	King Long	N45F	11/03	11/03	Michael Camenzuli, Marsascala
EBY 515	Bedford SB	Mulliner	B39F			
		Aquilina (11/71)	B40F	5/57	11/71	Joseph Grech, Mqabba
EBY 516	Bedford YRQ	Duple Dominant I	B45F	1/73	by 4/85	Francis Galea, Fgura
EBY 517	Bedford SBG	Brincat (1974)	B40F	1956	7/74	Paul Farrugia, Mqabba
EBY 518	Dodge f/c	Debono	B40F	by 12/65	by 7/73	Grezzju Grech, Zabbar
EBY 519	Bedford YRQ	Duple Dominant I	B45F	12/72	by 11/84	Alfred Mercieca, Marsa
EBY 520	Bristol LH6L	Eastern Coach Works	B45F	2/76	by 7/87	Victor Camilleri, Naxxar
EBY 521	Bedford YLQ	Duple Dominant II	B45F	5/77	by 8/86	Catherine Brincat, Qrendi
EBY 522	Bedford SB	Debono	B40F	1951	10/68	Spiridione Abela, Zejtun
EBY 523	Bristol LH6L	Eastern Coach Works	B45F	1/75	by 11/84	Carmel Borg, Zebbug
EBY 524	Bristol LH6L	Eastern Coach Works	B45F	12/74	by 9/85	Mario and Marisa Baldacchino, Ghaxaq

EBY 509	New
EBY 510	New
EBY 511	New
EBY 512	New
EBY 513, Y-0513, DDC104K	Morris and Evans, Llanfyllin: Begg, Middlesbrough
EBY 514	New
EBY 515, Y-0515, Y-0922, Y-1622, A-3343, 3343	Unscheduled bus, Malta (Y-0922): British Army 44 BS 62 with Mulliner B39F bodywork
EBY 516, Y-0516, TYG803L	New Enterprise, Tonbridge: Chatsworth Coaches, Pudsey
EBY 517, Y-0517, Y-1218, A-2425, 2425	Route Bus Malta, British Army 44BS63 with Mulliner B39F bodywork
EBY 518, Y-0518, Y-1219, A-2426, 2426	Cassar, Zabbar, Malta
EBY 519, Y-0519, MTX250L	Williams,Cwmdu, 17: Mainwaring, Gilfach Goch
EBY 520, Y-0520, OCA626P	Jones, Login: Crosville Motor Services, SLL626
EBY 521, Y-0521, OAN962R	Reliance, Newbury, 152
EBY 522, Y-0522, Y-1223, A-2435, 2435	Imported from Egypt in 10/68 via dealer in London
EBY 523, Y-0523, Y-1224, GLJ489N	Hants and Dorset, 3557
EBY 524, Y-0524, GLJ479N	Hants and Dorset, 3547

EBY 536, a Dodge forward control with Casha body, is parked outside the ATP headquarters in Blata l-Bajda in March 2009. Note its interesting route number, 46, which is one of few early morning short-workings that start at Mgarr instead of the normal terminus at Golden Bay of route 47.

EBY 525	AEC Swift 4MP2R1	Park Royal	B46F	10/70	1/82	Francis Galea, Fgura
EBY 526	King Long XMQ 6113GMC (m)	King Long	N45F	11/03	11/03	Joseph Bartolo, Mellieha
EBY 527	Bedford YLQ	Duple Dominant II	C45F	1/78	by 8/86	Joseph Caruana, Marsaxlokk
EBY 528	Bedford SB/Maltese Docks	Debono	B40F	9/69	9/69	Joseph Zammit, Qormi
EBY 529	Dodge f/c	Aquilina	B40F	11/54	11/54	Giovanni Camilleri, Mgarr
EBY 530	Maltese Docks/Leyland f/c	Barbara (1970)	B40F	1/59	1/59	Eleo Fenech, Gzira
EBY 531	King Long XMQ 6113GMC	King Long	N45F	1/04	1/04	Joseph Gatt, Zabbar
EBY 532	Dodge T110L f/c	Schembri (12/50); Debono (1/59)	B39F	1942	12/50	Joseph Zahra, Birzebbugia
EBY 533	King Long XMQ 6113GMC	King Long	N45F	11/03	11/03	Paul Borg, Mosta
EBY 534	AEC Swift 4MP2R2	MCW	B46F	4/71	10/81	Mario Attard, Zebbug
EBY 535	BMC Falcon	BMC	N45F	7/03	7/03	Anthony Saliba, Zabbar
EBY 536	Dodge f/c	Casha	B40F	by 12/59	by12/59	Felix Fenech, Mosta
EBY 537	Ford Thames ET7 n/c	Micallef	B36C	by 12/52	2/53	Anthony Falzon, Qormi
EBY 538	King Long XMQ 6113GMC	King Long	N45F	2/04	2/04	Paul & Rocco (Marsa Ltd), Marsa
EBY 539	Bedford YRQ	Plaxton Elite III	B45F	7/75	by 4/85	Mario Cassar, Birzebbugia
EBY 540	Bedford SL (Maltese)	Desono	B40F	12/67	12/67	Raymond Demanuele, Pieta
EBY 541	AEC Swift 4MP2R1	Park Royal	B46F	10/70	12/80	Paul & Rocco (Marsa Ltd), Marsa

EBY 525, Y-0525, Y-1226, A-2442, EGN237J	London Transport, SMS237
EBY 526	New
EBY 527, Y-0527, SJR307S	Tompkin, Countesthorpe: Craiggs, Amble
EBY 528, Y-0528, Y-1229, A-2458, 2458	New
EBY 529, Y-0529, Y-1230, A-2472, 2472	New
EBY 530, Y-0530, Y-1231, A-2484, 2484	New
EBY 531	New
EBY 532, Y-0532, Y-0859, Y-1559, A-5567, 5567	Route bus, Malta: War Department lorry, N160
EBY 533	New
EBY 534, Y-0534, Y-1235, A-2492, EGN572J	London Transport, SMS572
EBY 535	New
EBY 536, Y-0536, Y-1237, A-2500, 2500	Route bus by 1954 on Dodge truck chassis rebuilt to f/c 1960s
EBY 537, Y-0537, Y-1238, A-2501, 2501	New, as chassis
EBY 538	New
EBY 539, Y-0539, JNK559N	Plumpton Coaches: Wootten, London SE13
EBY 540, Y-0540, Y-1241, A-2520, 2520	New
EBY 541, Y-0541, Y-1242, A-2525, EGN247J	London Transport, SMS247

The Cirkewwa terminus finds EBY 556, a Bedford SL with Aquilina bodywork, on lay-over during a late August afternoon journey. Notice the 2571 fleet number just below the radiator grills. This is another characteristic of Maltese buses and recalls the registration of the original bus holding this licence.

EBY 542	BMC Falcon	BMC	N45F	7/03	7/03	Owen Spiteri, Mqabba
EBY 543	Leyland Leopard PSU4C/4R	Plaxton Derwent	B46F	4/76	1989	Mario Joseph Grech, Qormi
EBY 544	Bristol LH6L	Eastern Coach Works	B45F	12/74	by 8/86	Saviour Camilleri, Naxxar
EBY 545	Bedford YRQ	Duple Dominant I	C45F	7/74	4/86	Joseph Mifsud, Naxxar
EBY 546	Leyland Beaver	Grech	B46F	1965	by 11/89	Nazareno Agius, Zebbug
EBY 547	King Long XMQ 6113GMC	King Long	N45F	10/03	10/03	Silvio and Albert Camilleri, Naxxar
EBY 548	King Long XMQ 6113GM (m)	King Long	N45F	12/04	2/08	Martin Chetcuti, Mgarr
EBY 549	Bedford QL	Brincat	B40F	by 12/59	by 12/59	Sean Grech, Naxxar
EBY 550	Bedford YRQ	Plaxton Panorama Elite II	B45F	2/72	by 11/84	Paul & Rocco (Marsa Ltd), Marsa
EBY 551	Maltese f/c	Zammit (1969); Ciantar (1982)	B45F	9/69	9/69	Carmel Cutajar, Paola
EBY 552	Ford V8 f/c Sultana (10/47): Tonna conversion (1962)		B40F	8/39	8/39	Nicholas Mifsud, Qormi
EBY 553	King Long XMQ 6113GM	King Long	N45F	5/03	5/03	Tal-Linja Co. Ltd, Naxxar
EBY 554	AEC Swift 4MP2R	Marshall	B46F	4/70	11/81	Alex Pulo, Paola
EBY 555	AEC Mercury	Brincat	B40F	1957	11/67	George Casha, Qormi
EBY 556	Bedford SL (Maltese)	Aquilina	B40F	9/63	9/63	Carmel Farrugia, Tarxien
EBY 557	Dodge T110L f/c	Casha (8/53); Vella (1965)	B45F	1942	8/53	Mario Farrugia, Mqabba
EBY 558	Bedford YLQ	Plaxton Supreme III	B45F	11/76	by 10/86	Paul & Rocco (Marsa Ltd), Marsa

EBY 542	New
EBY 543, Y-0543, LUG524P	Wealden, Nettleshead: West Yorkshire PTE, 8524
EBY 544, Y-0544, GDV460N	Western National, 1611
EBY 545, Y-0545, STW705M	Ron's, Ashington: Freeman, Loughton
EBY 546, Y-0546, KNP469D	Lorry chassis imported from Everton of Droitwich in 11/81; body not complete until 1989
EBY 547	New
EBY 548	New *(note its delayed entry into service)*
EBY 549, Y-0549, Y-1250, A-2550, 2550	History unknown
EBY 550, Y-0550, Y-1251, WCF539K	Morley, West Row
EBY 551, Y-0551, Y-1252, A-2554, 2554	New, on imported West German chassis
EBY 552, Y-0552, Y-1253, A-2558, 2558	New, via Gasan, then as truck 534 during WW2
EBY 553	New
EBY 554, Y-0554, Y-0633, Y-1334, A-2881, AML16H	Route bus, Malta: London Transport, SM16
EBY 555, Y-0555, Y-1256, A-2570, 2570, 4592	Lorry chassis of 1957 vintage, released from customs in 4/64
EBY 556, Y-0556, Y-1257, A-2571, 2571	New
EBY 557, Y-0557	Dodge chassis new in 1942
EBY 558, Y-0558, NVA45R	Miller, Foxton

August 2008 finds AEC Swift EBY 564 it Cirkewwa. It was soon to depart on the long journey to Sliema Ferries on route 645. The Paradise Bay Hotel is in the background.

EBY 559	AEC Mercury	Farrugia	B40F	12/63	7/64	Roderick Muscat, Marsaxlokk
EBY 560	Bedford SLD	Gauci	B40F	1955	1/67	Joseph Camilleri, Rabat
EBY 561	MAN 14.220	Scarnif	N45F	12/04	12/04	Paul & Rocco (Marsa Ltd), Marsa
EBY 562	BMC Falcon	BMC	N45F	4/03	4/03	Horace Vella, Qormi
EBY 563	Bedford YRQ	Plaxton Elite Express II	BC45F	5/72	by 7/88	John Ciappara, Qormi
EBY 564	AEC Swift 4MP2R1	Park Royal/Busuttil (2001)	B46F	10/70	10/81	Paul & Rocco (Marsa Ltd), Marsa
EBY 565	Maltese f/c (Baileys)	Gauci	B40F	12/62	12/62	George Valletta, Qormi
EBY 566	Bedford YLQ	Duple Dominant II	BC45F	4/77	5/85	Charlton Caruana, Marsaxlokk
EBY 567	AEC Reliance 2MU3RA	Grech	B45F	1959	6/82	John Camilleri, Naxxar
EBY 568	Ford R1014	Duple Dominant IA	C45F	7/78	4/86	Paul & Rocco (Marsa Ltd), Marsa
EBY 569	Ford 510E	Zammit	B41F	8/63	8/69	Joseph Grech, Mqabba
EBY 570	BMC Falcon	BMC	N45F	5/03	5/03	Saviour Zahra, Rabat
EBY 571	Bedford QL	Barbara	B40F	by 6/60	9/62	Rodianne Sultana, Santa Venera
EBY 572	King Long XMQ 6113GMC	King Long	N45F	10/03	10/03	Joseph Grech, Mqabba
EBY 573	Bedford YRQ	Plaxton Panorama Elite II Scarnif	BC45F	10/73	5/85	Joseph Garcia, Fgura
EBY 574	Bedford YRQ	Plaxton Elite II	BC45F	1971	5/85	Rodianne Sultana, Santa Venera

EBY 559, Y-0559, Y-0900, Y-1600, A-2899, 2899	Unscheduled bus, Malta (Y-0900): originally a lorry
EBY 560, Y-0560, Y-1261, A-2853, 2853	Chassis imported 8/64 from lorry 115GMY
EBY 561	New
EBY 562	New
EBY 563, Y-0563, CJR823K	Woodcock, Buxton: Craiggs, Amble
EBY 564, Y-0564, Y-1265, A-2589, EGN268J	London Transport, SMS268
EBY 565, Y-0565, Y-1266, A-2590, 2590	New on locally built chassis constructed in 10/61
EBY 566, Y-0566, RAW14R	Evans, Penrhyncoch: Whittle, Highley, 14
EBY 567, Y-0567, Y-0861, Y-1561, 9191NW	Unscheduled bus, Malta (Y-0861): Cleveland Transit; Wallace Arnold *(chassis exported to Malta in 12/75)*
EBY 568, Y-0568, NJS622S	Pemberton, Upton: Newton, Dingwall
EBY 569, Y-0569, Y-1270, A-2595, 2595	Fenech, Gharghur, Malta: imported Ford chassis, bodied 8/89
EBY 570	New
EBY 571, Y-0571, Y-1272, A-2599, 2599	Bedford QL truck, 23403
EBY 572	New
EBY 573, Y-0573, OPT730M	Ross, Cotgrave: Armstrong, Ebchester
EBY 574, Y-0574, URO848J	Stored for many years and EBY574 has never been carried

EBY 586 is a Bedford SB of 1952 vintage with Casha bodywork. It is working route 47 to Golden Bay and is seen in the Ta' Zoqqrija suburb of Mosta.

EBY 575	Bedford SL or SB (Maltese)	Schembri	B40F	1/70	10/72	Carmel Agius, Paola
EBY 576	AEC Swift 4MP2R	Park Royal	B46F	4/70	11/81	Andrew Garnisi, Luqa
EBY 577	Bedford YLQ	Plaxton Supreme Express III	C45F	5/77	5/85	Paul & Rocco (Marsa Ltd), Marsa
EBY 578	King Long XMQ 6113GM (m)	King Long	N45F	11/03	11/03	Rodianne Sultana, Santa Venera
EBY 579	Bedford SB8	Sammut	B40F	1/59	1/59	Maria Antonia Zammit, Qormi
EBY 580	Bedford YRQ	Plaxton Panorama Elite III	BC45F	5/73	by 11/84	Clive Schembri, Zebbug
EBY 581	King Long XMQ 6113GMC	King Long	N45F	7/04	7/04	Maria Antonia Zammit, Qormi
EBY 582	King Long XMQ 6113GMC	King Long	N45F	10/03	10/03	Maria Borg, Mellieha
EBY 583	Bedford YLQ	Plaxton Supreme Elite III	BC45F	5/77	by 2/88	Mario Bonavia, Naxxar
EBY 584	Bedford SL	Falzon	B40F	1956	11/65	Charles Cutajar, Zabbar
EBY 585	Bedford YRQ	Duple Dominant I	C45F	9/74	by 1/86	Spiridione Pulo, Santa Venera
EBY 586	Bedford SB	Casha	B40F	1952	by 8/74	Martin Galea, Rabat
EBY 587	Bedford YRQ	Plaxton Panorama Elite III	BC45F	1/73	by 11/84	Winstin Muscat, St Paul's Bay
EBY 588	King Long XMQ 6113GMC	King Long	N45F	6/03	6/03	Silvio Borg, Mellieha
EBY 589	AEC Swift 4MP2R1	Park Royal	B46F	1/71	by 10/81	Paul & Rocco (Marsa Ltd), Marsa
EBY 590	Maltese Docks/AEC f/c	Debono	B40F	10/68	10/68	Gaetano Fenech, Zejtun
EBY 591	Bedford YLQ	Plaxton Supreme III	C45F	3/77	by 10/87	Victor Spiteri, Birkirkara

EBY 575, Y-0575, Y-1276, A-2620, 2620	Briffa, San Gwann, imported from GB 1968 assembled locally and bodied by 10/72
EBY 576, Y-0576, Y-1277, A-2621, AML60H	London Transport, SMD60
EBY 577, Y-0577, OHY790R	Thomas of Barry
EBY 578	New
EBY 579, Y-0579, Y-1280, A-2630, 2630	New
EBY 580, Y-0580, Y-1281, XWW791L	Morley, West Row: Central Motors, Ripponden
EBY 581	New
EBY 582	New
EBY 583, Y-0583, WTU123R	Brownrigg & Cook, Egremont: Bullock, Cheadle
EBY 584, Y-0584, Y-1285, A-2676, 2676	Second-hand Bedford chassis imported from UK 10/63 but not bodied until 11/65
EBY 585, Y-0585, RAL968N	Cooper, Killamarsh: Kettlewell, Retford
EBY 586, Y-0586, Y-1287, A-2705, 2705	Ministry of Supply
EBY 587, Y-0587, Y-1288, CNT265L	Williams, Cwmdu: Hampson, Oswestry, 34
EBY 588	New
EBY 589, Y-0589, Y-1290, EGN292J	London Transport, SMS292
EBY 590, Y-0590, Y-1291, A-2714, 2714	New
EBY 591, Y-0591, PET214R	Halliday Hartle, Buxton

One of the Bristol LHs now in Malta is EBY 592, which was latterly OCA635P (SLL635) in the Crosville fleet. It is seen at the Mater Dei Hospital on route 675 having arrived from St Andrew's.

EBY 592	Bristol LH6L	Eastern Coach Works	B45F	3/76	by 10/86	Natalie Camilleri, Naxxar
EBY 593	Bedford YRQ	Duple Dominant I	BC45F	3/75	by 9/85	Carmelo Muscat, Naxxar
EBY 594	Bedford YRQ	Duple Dominant I	C45F	9/74	1/85	Pauline Ellul, Tarxien
EBY 595	Ford R1014	Duple Dominant Bus	B46F	12/78	8/91	Owen Spiteri, Mqabba
EBY 596	Bedford YRQ	Duple Dominant Express	C45F	10/75	by 11/84	Michael Camenzuli, Marsascala
EBY 597	Bedford YRQ	Plaxton Panorama Elite III	BC45F	4/75	by 8/86	Grazio Borg, Zabbar
EBY 598	Volvo B7RLE	Saracakis	N45F	7/03	7/03	Philip Caruana, Zebbug
EBY 599	Bedford YRQ	Plaxton Panorama Elite III	BC49F	9/74	by 1/86	Charlie Abela, Zejtun
EBY 600	King Long XMQ 6113GMC	King Long	N45F	11/03	11/03	Filippa d'Amato, Zebbug
EBY 601	King Long XMQ 6113GMC	King Long	N45F	11/03	11/03	Rodianne Sultana, San Gwann
EBY 602	BMC Falcon	BMC	N45F	6/03	6/03	Joseph Saliba, Zabbar
EBY 603	AEC Matador	Gauci	B40F	1951	8/65	Michael Spiteri, Fgura
EBY 604	Ford R1014	Plaxton Panorama Elite III	C45F	3/75	by 4/85	CMS Transport, (Stephen Cilia), Gharghur
EBY 605	Bedford YLQ	Plaxton Supreme III	BC45F	3/77	by 2/88	Francis Galea, Fgura
EBY 606	Bedford YLQ	Duple Dominant II	BC45F	2/79	by 5/95	Francis Galea, Fgura
EBY 607	Bedford YRQ	Plaxton Panorama Elite III	BC45F	10/74	by 4/90	Joseph Mifsud, Mellieha

EBY 592, Y-0592, OCA635P	Jones, Login: Crosville Motor Services, SLL635
EBY 593, Y-0593, HNT841N	Day, Abertillery: M&M, Highley, 33
EBY 594, Y-0594, SMR662N	Stanfield, Figheldean
EBY 595, Y-0595, COO242T	Suffolk County Council, Ipswich: Ford, Warley
EBY 596, Y-0596, Y-1297, KBX39P	Davies Bros., Pencader, 99
EBY 597, Y-0597, HMV645N	Ron's, Ashington: Bexleyheath Transport, 105
EBY 598	New
EBY 599, Y-0599, SJR426N	Holmes, Clay Cross: Moordale, Newcastle
EBY 600	New
EBY 601	New
EBY 602	New
EBY 603, Y-0603, Y-1304, A-2814, 2814	RAF AEC lorry. Chassis purchased 9/62
EBY 604, Y-0604, JDS644N	Taylor, Derby: Northern Roadways, Glasgow
EBY 605, Y-0605, TWT504R	Evans, Tregaron: Anderton, Keighley
EBY 606, Y-0606, DTM958T	Williams, Brecon: Tate, Markyate
EBY 607, Y-0607, Y-0431, Y-1132, UUP3N	Route bus, Malta: Armstrong of Ebchester

EBY 620 is a Leyland Royal Tiger that was originally delivered to Bournemouth Corporation in 1954. It now carries locally built Caruana bodywork of 1979 vintage. Route 49 links Qawra and Bugibba with the capital, and the bus is seen glinting in the early morning sun at Qawra bus terminus.

EBY 608	Bedford YRQ	Plaxton Panorama Elite Exp III	C45F	8/75	by 10/87	Charles Schembri, Qormi
EBY 609	Bedford YLQ	Plaxton Supreme III	C45F	8/76	by 10/87	Grazio Borg, Zabbar
EBY 610	Bedford YLQ	Duple Dominant I	BC45F	5/76	4/86	Martin Borg, Mellieha
EBY 611	Bristol LH6L	Plaxton Supreme II Express	BC45F	11/77	by 5/90	Joseph Xuereb, Mellieha
EBY 612	AEC f/c	Debono (1971)	B44F	1958	12/71	Charles Farrugia, Mqabba
EBY 613	King Long XMQ 6113GMC	King Long	N45F	11/03	11/03	Saviour Zammit, Tarxien
EBY 614	Bedford YRQ	Duple Dominant I	BC45F	1/75	by 7/87	Paul & Rocco (Marsa Ltd), Marsa
EBY 615	Bedford YRQ	Plaxton Panorama Elite Exp III	C45F	5/76	by 1/85	Martin Bartolo, Mosta
EBY 616	Maltese Docks f/c	Barbara	B40F	3/70	3/70	Marco Magro, Zurrieq
EBY 617	Bedford SB3	Debono	B42F	1957	10/58	David Vella, San Gwann
EBY 618	Bedford SL (Zammit)	Barbara	B43F	3/71	3/71	Rodianne Sultana, Santa Venera
EBY 619	Ford R1014	Willowbrook	B46F	6/78	by 2/92	Andrew Abela, Zejtun
EBY 620	Leyland Royal Tiger PSU1/13	Caruana (1979)	B45F	1954	by 6/80	Charles Calleja, Naxxar
EBY 621	MCII	MCII	N45F	6/03	6/03	Jane Farrugia, Tarxien
EBY 622	Leyland Royal Tiger PSU1/16	Aquilina (1975)	B44F	6/53	8/75	Paul & Rocco (Marsa Ltd), Marsa
EBY 623	AEC Swift 4MP2R	Marshall	B46F	10/70	11/82	Francis Galea, Fgura

EBY 608, Y-0608, KDF856P — Pulham, Bourton-on-the-Water

EBY 609, Y-0609, SUP436R — Chivers, Elstead: Armstrong, Ebchester

EBY 610, Y-0610, ODU251P — Sayer, Ipswich: Wainfleet, Nuneaton

EBY 611, Y-0611, RDE298S — Williams, Cwmdu: Richards, Moylgrove

EBY 612, Y-0612, Y-0915, Y-1615, A-208, 208 — Unscheduled bus, Malta (Y-0915); imported AEC lorry chassis

EBY 613 — New

EBY 614, Y-0614, GUX949N — Rees, Llanelly Hill: Corvedale, Ludlow, 45

EBY 615, Y-0615, Y-1316, NTO34P — Leiston Motor Hire: Skills, Nottingham, 29

EBY 616, Y-0616, Y-1317, A-2838, 2838 — New

EBY 617, Y-0617, Y-1318, A-2839, 2839 — Bedford SB chassis imported in 12/57. This vehicle was re-instated as a route bus in late 2007

EBY 618, Y-0618, Y-1319, A-2840, 2840 — Locally assembled Bedford SL chassis Xuereb, Mellieha, Malta

EBY 619, Y-0619, XKX856S — Hampton, London SE1: Whyte, Colnbrooke, 56

EBY 620, Y-0620, Y-0856, Y-1556, A-3357, 3357, NLJ270 — Unscheduled bus, Malta (Y-0856): Bournemouth, 92

EBY 621 — New

EBY 622, Y-0622, Y-0883, Y-1583, A-5571, 5571, FCK433 — Zarb Coaches, Malta: Ribble, 933

EBY 623, LCY 907, Y-0623, Y-1324, EGN274J — London Transport, SMS274

EBY 637 is a Leyland Tiger Cub with East Lancs bodywork which was originally in the Blackburn Transport fleet. It now boasts a Bristol LH front and is seen at the bus terminus in Naxxar in September 2009, with flags and streamers clearly in view in advance of the town's *Festa* taking place the following weekend.

EBY 624	Bedford YRQ	Duple Dominant Express	BC45F	8/75	by 2/88	Marija Buttigieg, Zabbar
EBY 625	AEC Swift 4MP2R	Marshall	B46F	1/70	6/81	Paul Spiteri, Qrendi
EBY 626	Bedford YLQ	Plaxton Panorama Elite III Exp	BC45F	7/77	by 4/85	Pacifico Scerri, Mosta
EBY 627	BMC Falcon	BMC	N45F	5/03	5/03	Carmelo Cassar, Mqabba
EBY 628	Maltese Docks/AEC f/c	Barbara	B40F	1964	1964	Aneuren Meli, Tarxien
EBY 629	Commer Avenger IV	Debono	BC--F	1960	2/69	Rodianne Sultana, Santa Venera
EBY 630	Bedford YLQ	Plaxton Supreme III	BC45F	8/78	8/89	Mario Farrugia, Attard
EBY 631	Dodge T110 f/c	Casha	B39F	by 12/59	by 1/86	Rennie Bonnici, Ghaxaq
EBY 632	Bedford SL	Zammit (probably)	B40F	1956	5/66	Joseph Demanuele, Hamrun
EBY 633	AEC Swift 4MP2R2	MCW	B46F	5/71	by 5/92	CMS Transport (Stephen Cilia), Gharghur
EBY 634	AEC Swift 4MP2RI	Marshall	B46F	3/71	by 8/81	Paul Zammit, Tarxien
EBY 635	King Long XMQ6113GMC	King Long	N45F	10/03	10/03	Thomas Borg, Mellieha
EBY 636	Ford V8 n/c	BMC/Gasan	B40F	7/49	1/51	Joseph Borg, Zabbar
EBY 637	Leyland PSUC1/13	East Lancs	B46F	1967	by 8/81	Albert Camilleri, Naxxar
EBY 638	King Long XMQ 6113GMC	King Long	N45F	11/03	11/03	Raymond Aquilina, Mellieha

EBY 624, Y-0624, JDK500P	Ouse Valley, Goldington: Bywater, Rochdale
EBY 625, Y-0625, Y-1326, A-2862, AML8H	London Transport, SM 8
EBY 626, Y-0626, Y-1327, CAA840R	Razey, Thruxton
EBY 627	New
EBY 628, Y-0628, Y-1329, A-2850, 2850	History unknown
EBY 629, Y-0629, Y-0896, Y-1596, A-3334, 3334	Unscheduled bus, Malta (Y-0896). Chassis imported 1960 but not bodied until 1969; currently stored with body reduced to its frame
EBY 630, Y-0630, UFT912T	Briggs, Swansea: Curtis, Dudley
EBY 631, Y-0631, Y-1332, A-2878, 2878, 2128	Unscheduled bus, Malta
EBY 632, Y-0632, Y-1333, A-2879, 2879	Bedford SL lorry chassis imported from London
EBY 633. Y-0633, Y-0554, Y-1255, A-2566, EGN603J	Route bus, Malta: London Transport, SMS 603
EBY 634, Y-0634, Y-1335, A-2882, EGN212J	London Transport, SMS 212
EBY 635	New
EBY 636, Y-0636, Y-1337, A-2890, 2890	US Ford chassis imported in 7/49; in service 1/51
EBY 637, Y-0637, Y-1338, A-2895, GBV13E	Blackburn, 13
EBY 638	New

FBY 648, a Ford ET7 f/c bus with Aquilina bodywork is now over fifty years old. It was pictured at the terminus of the Hospital/University routes in Tal Qroqq one lunchtime. Route 450 serves Birkirkara, Mosta, St Paul's Bay and Mellieha on its way to the Cirkewwa terminus.

EBY 639	MAN 14.220	Scarnif	N45F	3/03	3/03	John Caruana, Mosta
EBY 640	King Long XMQ 6113GMC	King Long	N45F	12/03	12/03	Donald Borg, Mosta
FBY 641	AEC Swift 4MP2R1	Marshall	B46F	9/71	6/81	Tal-Linja Co. Ltd, Naxxar
FBY 642	AEC Swift 4MP2R1	Park Royal	B46F	10/70	11/81	Francis Galea, Fgura
FBY 643	Bedford YLQ	Plaxton Supreme III	C45F	3/77	by 9/85	Joseph and Stephen Scerri, Mosta
FBY 644	Bedford SB	Brincat	B40F	1966	1966	Francis Galea, Fgura
FBY 645	Bedford SB	Aquilina (1970)	B40F	1951	1952	Albert Sammut, San Gwann
FBY 646	Bedford YMP	Marshall	B45F	11/83	12/01	Francis Galea, Fgura
FBY 647	King Long XMQ 6113GMC	King Long	N45F	11/03	11/03	Emmanuele Cortis, Naxxar
FBY 648	Ford ET7 f/c	Aquilina	B40F	1/57	1/57	Anthony Schembri, Qormi
FBY 649	King Long XMQ 6113GMC	King Long	N45F	6/03	6/03	Charles Seychell, Fgura
FBY 650	Bedford YLQ	Duple Dominant I	BC45F	5/76	by 9/85	Victor Camilleri, Naxxar
FBY 651	Bedford YRQ	Plaxton Panorama Elite II	BC45F	3/71	by 1/85	Alfred Mercieca, Marsa
FBY 652	Bedford YRQ	Duple Dominant I	C45F	5/75	by 5/95	Franco Camilleri, Naxxar
FBY 653	King Long XMQ 6113GMC	King Long	N45F	10/03	10/03	Alexander Schembri, Qormi
FBY 654	Bedford YRQ	Plaxton Panorama Elite II	C45F	5/72	by 9/85	Albert Sammut, San Gwann
FBY 655	Bedford YRQ	Plaxton Panorama Elite II	BC45F	4/72	by 4/90	Francis Galea, Fgura

EBY 639 — New
EBY 640 — New
FBY 641, Y-0641, Y-1342, A-2925, EGN202J — London Transport, SMS202
FBY 642, Y-0642, Y-1343, EGN243J — London Transport, SMS243
FBY 643, Y-0643, MPX7R — Bell, Winterslow: Bicknell, Godalming
FBY 644, Y-0644, Y-1345, A-2958, 2958 — Locally assembled chassis *(records reveal no more history)*
FBY 645, Y-0645, Y-1346, A-2980, 2980 — Ministry of Supply, Bedford SB with Mulliner B31FA bodywork, later with RAF registration 27AC58
FBY 646, A203LCL — Lewis, Llanrhystyd: Norfolk County Council Education Department
FBY 647 — New
FBY 648, Y-0648, Y-0867, Y-1567, A-5575, 5575 — Unscheduled bus, Malta (Y-0867)
FBY 649 — New
FBY 650, Y-0650, MTX661P — Rees, Llanelly Hill
FBY 651, Y-0651, HWW 60J — Williams, Cwmdu: Steel, Addingham
FBY 652, Y-0652, JAP441N — Camden, Sevenoaks: Alpha, Brighton
FBY 653 — New
FBY 654, Y-0654, RPR740K — Adams, Handley: Rendell, Parkstone
FBY 655, Y-0655, Y-0657, UNU178K — Halliday Hartle, Buxton

FBY 667 is a Bedford SB with Barbara bodywork and has been in service as a route bus since 1964. It is seen at Qawra bus terminus in September 2009 and unusually sports a destination board in the windscreen for route 580 that serves the Mater Dei Hospital and University complex at Tal Qroqq.

Reg	Chassis	Body	Seating	Date	Date	Owner
FBY 656	AEC Swift 4MP2R	Marshall	B46F	2/70	4/81	Anthony Darmanin, Zabbar
FBY 657	Bedford YMQ	Marshall	B49F	12/81	by 2/92	Andrew Abela, Zejtun
FBY 658	Leyland Super Comet 14SC	Aquilina (1967)	B40F	1963	10/67	Silvio Schembri, Qormi
FBY 659	AEC Mustang GM6RH	Aquilina (1968)	B39F	1958	8/68	Joseph Grech, Mqabba
FBY 660	Bedford YLQ	Duple Dominant	BC45F	10/78	by 1/87	Franco Caruana, Zabbar
FBY 661	Willys Six Model 265 n/c	Brincat (1949)	B34F	1933	1934	Rosaria Falzon, Fgura
FBY 662	AEC Swift 4MP2R	Marshall	B46F	2/70	6/81	Francis Galea, Fgura
FBY 663	Bedford QL	Barbara (1963)	B40F	by 12/44	2/63	Consiglio Gatt, Qormi
FBY 664	BMC Falcon	BMC	N45F	7/03	7/03	Joseph Scerri, Mosta
FBY 665	AEC Swift 4MP2R5	MCW	B46F	1/72	7/81	Paul & Rocco (Marsa Ltd), Marsa
FBY 666	Leyland Super Comet 14SC	Barbara	B40F	5/64	5/70	Raymond Borg, Mellieha
FBY 667	Bedford SB	Barbara	B45F	1955	5/64	Anthony Xuereb, Naxxar
FBY 668	This registration is currently not in use					George Micallef, Qormi
FBY 669	Bedford SB	Brincat	B41F	1959	1959	Noel Cassar, Zabbar
FBY 670	Bedford YLQ	Duple Dominant II	C45F	7/77	by 7/87	Emmanuele Valletta, Qormi
FBY 671	BMC Falcon	BMC	N45F	6/03	6/03	Z.S. Co. Ltd, Naxxar
FBY 672	Maltese f/c (Zammit)	Brincat	B40F	2/68	3/69	Paul & Rocco (Marsa Ltd), Marsa

FBY 656, Y-0656, Y-0321, Y-1022, A-0358, AML14H — Route bus, Malta: London Transport, SM14
FBY 657, Y-0657, WVG788X — Norfolk County Council Education Department
FBY 658, Y-0658, Y-1359, A-3063, 3063 — Lorry chassis imported in 10/66
FBY 659, Y-0659, Y-0884, Y-1584, A-2999, 2999 — Zarb Coaches, Mosta (Y-0884);
AEC Mustang lorry chassis, new in 1958 and imported in 3/65
FBY 660, Y-0660, ANR743T — Fourways, Leeds
FBY 661, Y-0661, Y-0857, Y-1557, A-3036, 3036, 2219 — Unscheduled bus, Malta (Y-0857)
FBY 662, Y-0662, Y-1363, A-3075, AML33H — London Transport, SM33
FBY 663, Y-0663, Y-0851, Y-1551, A-896, 896 — Unscheduled bus, Malta (Y-0851) RAF 4x4 lorry chassis imported 6/59
FBY 664 — New
FBY 665, Y-0665, Y-1366, A-3081, JGF810K — London Transport, SMS810, currently in store
FBY 666, Y-0666, Y-1367, A-3082, 3082 — New on a chassis imported from GB in 5/64
FBY 667, Y-0667, Y-1368, A-3083, 3083 — RN (HM Dockyard, Malta) Original body scrapped in 12/62 and rebodied 4/64
FBY 668 —
FBY 669, Y-0669, Y-0911, Y-1611, A-2674, 2674 — Unscheduled bus, Malta (Y-0911)
FBY 670, Y-0670, RAW 44R — Rees, Llanelly Hill: Corvedale, Ludlow, 44
FBY 671 — New
FBY 672, Y-0672, Y-1373, A-3098, 3098 — Chassis frames imported in 7/65 by Zammit and assembled in 2/68

This AEC Reliance with an Aquilina body, FBY 676, is on lay-over close to the main entrance of the Mater Dei Hospital during a dull afternoon in March 2009.

FBY 673	Bedford YLQ	Duple Dominant I	BC45F	4/77	4/86	Noel Desira, Zabbar
FBY 674	AEC Reliance 2MU3RA	Ciantar (1979)	B45F	1961	by 6/81	John Spiteri, Qormi
FBY 675	AEC Swift 4MP2R	Marshall	B46F	2/70	2/81	Joseph Muscat, Siggiewi
FBY 676	AEC Reliance MV3RU	Aquilina (C36F in 1975)	B45F	6/57	7/75	Paul Camilleri, Naxxar
FBY 677	King Long XMQ6113GMC	King Long	N45F	11/03	11/03	Alfred Mercieca, Marsa
FBY 678	King Long XMQ 6113GMC	King Long	N45F	10/03	10/03	Francis Caruana, Fgura
FBY 679	King Long XMQ 6113GMC	King Long	N45F	3/04	3/04	Marija Buttigieg, Zabbar
FBY 680	Dodge T110L f/c	Sammut	B40F	by 2/54	by 2/54	Christopher Buttigieg, Zabbar
FBY 681	Bedford QLD	Debono	B40F	1945	11/58	Paul & Rocco (Marsa Ltd), Marsa
FBY 682	Bedford YRQ	Duple Dominant I	BC45F	2/73	by 4/85	Anthony Schembri, Qormi
FBY 683	Bedford YMT	Plaxton Supreme III	BC47F	4/77	by 10/86	Francis Galea, Fgura
FBY 684	Bedford QL	Casha	B40F	by 11/57	5/62	Mario Sultana, Cospicua
FBY 685	Maltese Docks f/c	Debono	B45F	1968	1968	Raymond Gialanze, Zabbar
FBY 686	King Long XMQ 6113GMC	King Long	N45F	11/04	11/04	Raymond Borg, Mellieha
FBY 687	BMC Falcon	BMC	N45F	5/03	5/03	Anthony Cassar, Zabbar
FBY 688	BMC Falcon	BMC	N45F	5/03	5/03	Joseph Pace, Qrendi

FBY 673, Y-0673, OAN963R — Reliance, Newbury, 153
FBY 674, Y-0674, Y-0870, Y-1570, RDB847 — Zarb Coaches, Malta (Y-0870): North Western, 847
FBY 675, Y-0675, Y-1376, A-3114, AML11H — London Transport, SM 11
FBY 676, Y-0676, Y-0877, Y-1577, A-4398, 4398, YKR234 — Sultana, Gzira, Malta; Smith, Rochdale: Maidstone and District 3234; Chassis exported to Sultana in 5/75

FBY 677 — New
FBY 678 — New
FBY 679 — New
FBY 680, Y-0680, Y-1381, A-3135, 3135 — Agius, Qormi
FBY 681, Y-0681, Y-1382, A-3171, 3440 — Route bus, Malta
FBY 682, Y-0682, BUX239L — Glyn Williams, Lower Tumble: M & M, Cleobury Mortimer, 35
FBY 683, ACY 912, Y-0912, STA259R — Cancu Supreme, Malta (ACY 912): Snell, Newton Abbot
FBY 684, Y-0684, Y-1385, A-3175, 3175 — Falzon, Zabbar: chassis purchased at a War Department auction in 11/57

FBY 685, Y-0685, Y-0893, Y-1593 — Unscheduled bus, Malta (Y-0893)
FBY 686 — New
FBY 687 — New
FBY 688 — New

King Long FBY 697 carries a prominent advert for MacDonalds. On lay-over in the town centre at Qabbar, it was photographed only a few minutes later than the view of FBY 680 shown on page 61.

FBY 689	Bedford YLQ	Duple Dominant I	BC45F	2/77	by 10/86	Jesmar Desira, Zejtun
FBY 690	Bedford YRQ	Plaxton Panorama Elite III	BC45F	6/75	by 1/86	Manuel Cassar, Zabbar
FBY 691	Bedford YRQ	Plaxton Panorama Elite II	BC45F	4/71	by 4/85	Rodianne Sultana, Santa Venera
FBY 692	Bedford YRQ	Duple Dominant I	BC45F	1/74	by 3/86	John Abela, Zejtun
FBY 693	Bedford YRQ	Duple Dominant I	C45F	4/74	by 4/85	Tracy Azzopardi, San Gwann
FBY 694	Bedford SB5	Grech	B45F	1956	9/80	George Xerri, Marsa
FBY 695	Ford 6 f/c	Casha B34C (1960);				
		Barbara B36F (4/63)	B40F	12/45	7/59	Christopher Buttigieg, Zabbar
FBY 696	Bedford SL	Aquilina (1964)	B43F	1953	3/64	Natasha Sammut, Mosta
FBY 697	King Long XMQ 6113GMC	King Long	N45F	11/03	11/03	Angelo Sammut, Zebbug
FBY 698	Austin K5	Aquilina (1954)	B43F	1942	6/54	Albert Sammut, San Gwann
FBY 699	Dennis Lance SLF	Wright Pathfinder	N45F	4/94	1/08	Albert Camilleri, Naxxar
FBY 700	Bedford YRQ	Plaxton Panorama Elite III	BC45F	1/73	by 4/85	Angelo Muscat, Zebbug
FBY 701	GMC f/c	Brincat	B40F	by 12/45	4/61	Etienne Falzon, Siggiewi
FBY 702	Leyland Hippo Mark 1	Bonavia (1949); Cassar (2000)	B45F	by 1940	3/64	Emmanuele Cassar, Zabbar
FBY 703	Leyland Lynx	Leyland	N45F	2/90	8/07	Z. S. Co. Ltd, Naxxar
FBY 704	King Long XMQ 6113GMC	King Long	N45F	10/03	10/03	Marija Buttigieg, Zabbar
FBY 705	King Long XMQ 6113GMC	King Long	N45F	11/03	11/03	Silvio Muscat, Birkirkara
FBY 706	Bedford YRQ	Plaxton Panorama Elite III	BC45F	5/73	by 3/85	Nicholas Micallef, Qormi
FBY 707	Leyland Tiger Cub PSUC1/1	(under rebuild)	B(45)F	5/58	by 4/86	Rodianne Sultana, Santa Venera
FBY 708	Bedford YRQ	Marshall	B45F	7/75	by 6/89	Justin Borg, Attard
FBY 709	Bedford SB1	Barbara	B40F	12/59	12/59	Silvio Schembri, Qormi
FBY 710	King Long XQM 6113GMC	King Long	N45F	10/03	10/03	Paul Sant, Mosta
FBY 711	King Long XQM 6113GMC	King Long	N45F	9/04	9/04	Carmel Farrugia, Tarxien
FBY 712	Bedford SB	Schembri	B36F	12/56	12/56	Rodianne Sultana, Santa Venera
(Though withdrawn since 10/98 this vehicle still retains its registration)						
FBY 713	Dodge T110L f/c	Zammit	B40F	1945	1958	Joseph Demanuele, Hamrun
(Stored since 9/02)						
FBY 714	Bristol LH6L	Eastern Coach Works	B45F	1/75	1/85	Stephen Attard, Paola
FBY 715	Bedford YRQ	Duple Dominant I	BC45F	8/74	by 11/84	Rosaria Falzon, Fgura
FBY 716	Leyland Tiger TS7	Aquilina (1958)	B40F	by 12/39	11/58	Anthony Sant, Mellieha
FBY 717	Bedford YRQ	Plaxton Panorama Elite II	BC45F	11/71	by 4/85	Grezzju Borg, Zabbar
FBY 718	King Long XQM 6113GMC	King Long	N45F	11/03	11/03	Alfred Mercieca, Marsa
FBY 719	Bedford SB5	Brincat	B39F	1964	2/71	Francis Cutajar, Zabbar
FBY 720	Volvo B6-50	Alexander Dash	N43F	1/95	3/08	Carmel Farrugia, Tarxien

This Bedford YRQ with Marshall bodywork was originally in service with the Atomic Weapons Research Establishment in Aldermaston, though is seen here as FBY 708 at Qawra bus terminus with passengers waiting for it to leave for Valletta on route 159.

FBY 689, Y-0689, YAA260R	Marchwood, Totton
FBY 690, Y-0690, KAC440N	Letham, Dunfermline: Shaw, Coventry
FBY 691, Y-0691, URO849J	Morley, West Row: Wootten, London SE13
FBY 692, Y-0692, YAB600M	Down, Mary Tavy: Halford, Kempsey
FBY 693, Y-0693, VPF41M	Fale, Coombe Down: Safeguard, Guildford
FBY 694, Y-0694, Y-1395, A-3225, 3225	Second-hand chassis from Ministry of Supply
FBY 695, Y-0695, Y-1396, A-3230, 3230, 2484	Route bus Malta.
FBY 696, Y-0696, Y-1397, A-3231, 3231, 3132	Bedford SL chassis imported 4/63 from GB and bodied 3/64
FBY 697	New
FBY 698, Y-0698, Y-1399, A-3240, 3240	Austin K5 military lorry of 1942 vintage
FBY 699, L38WLH	UK North, Manchester: London Buses, LLW38
FBY 700, Y-0700, Y-1401, UWX113L	McLaughlin, Penwortham: Rhodes, Guiseley
FBY 701, Y-0701, Y-0855, Y-1555, A-7243, 7243	Unscheduled bus, Malta (Y-0855). WW2 military GMC lorry chassis rebuilt by Tanti with spares 6/59 and bodied in 4/61
FBY 702, Y-0702, Y-1403, A-3249, 3249	Royal Navy lorry, RN 28749
FBY 703, G221EOG	Travel West Midlands, 1221
FBY 704	New
FBY 705	New
FBY 706, Y-0706, Y-1407, PGW638L	Kirkbright, Colne: Grey Green, London N16
FBY 707, Y-0707, Y-0889, Y-1589, A-3073, 3073, NHE133	Aquilina, Msida: Yorkshire Traction Co., 1126
(As late as March 2002, this vehicle still carried its Y-0707 registration)	
FBY 708, Y-0708, HGM614N	Atomic Weapons Research Establishment, Aldermaston
FBY 709, Y-0709, Y-1410, A-3268, 3268	New on an imported chassis
FBY 710	New
FBY 711	New
FBY 712, Y-0712, Y0875, Y1575, A7207, 7207, 93	New
FBY 713, Y-0713, Y-1414, A3276, 3276	War Department lorry, 80YC85
FBY 714, Y-0714, Y-1415, GLJ485N	Hants and Dorset, 3553
FBY 715, FBY 780, FBY 715, Y-0715, RUX278N	Parfitt's, Rhymney Bridge: Corvedale, Ludlow, 48
FBY 716, Y-0716, Y-1417, A-3280, 3280, AWX763	Pennine of Gargrave
FBY 717, Y-0717, JTD615K	Morley, West Row: Cross and Sobey, St Anne's
FBY 718	New
FBY 719, Y-0719, Y-1420, A-3284, 3284	Bedford SB5 chassis imported from GB 1/68, bodied 2/71.
FBY 720, M125VAK	Stagecoach 30214: Yorkshire Terrier, Sheffield, 125

FBY 732 is a recent arrival on Malta. It is a Dennis Lance with Wright Pathfinder bodywork that was latterly with UK North in Manchester. It is parked across the road from the main lay-over area by St James's Bastion in Valletta with the Phoenicia Hotel in the background.

FBY	Make/model	Body	Seating			Operator
FBY 721	Maltese f/c (Baileys)	Brincat	B40F	3/62	3/62	Charlie Abela, Zejtun
FBY 722	Bedford YLQ	Marshall	B46F	10/80	by 5/92	Mario Abela, Zejtun
FBY 723	AEC Swift 4MP2R5	MCW	B46F	1/72	7/81	Mario Sultana, Cospicua
FBY 724	Ford R1014	Duple Dominant II	C45F	8/78	by 8/86	Antoine Sant, Mellieha
FBY 725	AEC Mercury GM4RH	Barbara	B40F	3/64	4/68	Nazzareno Abela, Zejtun
FBY 726	AEC Reliance MU3RV	Aquilina (1971)	B45F	1954	2/71	Victor Meli, Mosta
FBY 727	Volvo B6BLE	Scarnif	N45F	5/02	5/02	Saviour Vella, Zejtun
FBY 728	Bedford SB	Debono	B45F	1958	1958	Carmel Cutajar, Valletta
FBY 729	Bedford YLQ	Duple Dominant II	C45F	4/77	6/86	C.M.S. Transport Ltd (Stephen Cilia), Gharghur
FBY 730	AEC Mercury GM4RAE	Barbara	B45F	by 12/69	by 12/69	Auxilio Caruana, Mellieha
FBY 731	BMC Falcon	BMC	N45F	4/03	4/03	Horace Vella, Qormi
FBY 732	Dennis Lance SLF	Wright Pathfinder	N45F	6/94	3/08	Mario Abela, Zejtun
FBY 733	Maltese (Baileys)	Brincat	BC36F	1983	1993	Rosaria Falzon, Fgura
FBY 734	*This vehicle was destroyed by fire in January 2010*					Z S Co., Naxxar
FBY 735	Bedford YRQ	Duple Dominant I	BC45F	9/73	1/85	Z S Co., Naxxar

FBY 721, Y-0721, Y-1422, A-3286, 3286 — New chassis frame, fitted out using Bedford spares
FBY 722, Y-0722, PEX738W — Norfolk County Council Education Department
FBY 723, Y-0723, Y-1424, A-3288, JGF803K — London Transport, SMS803
FBY 724, Y-0724, LHO418T — Marchwood, Totton
FBY 725, Y-0725, Y-1426, A-3290, 3290 — AEC Mercury chassis imported in 3/64
FBY 726, FBY 772, FBY 726, Y-0726, Y-0876, Y-1576, A-0764, 764, TUA17 — Route bus, Malta: Wallace Arnold Tours, Leeds exported in 8/68
FBY 727 — New
FBY 728, Y-0728, Y-1429, A-3293, 3293 — New
FBY 729, Y-0729, RAW27R — Daniel, Cardigan: Whittle, Highley, 27
FBY 730, Y-0730, Y-1431, A-3297, 602 — Route bus, Malta
FBY 731 — New
FBY 732, L32WLH — UK North, Manchester: London Buses, LLW32
FBY 733, Y-0733 — Chassis built in 1983; bodywork not finished until 1993. Never used in public service

FBY 734 — -
FBY 735, Y-0735, OOT267M — Pulfrey, Great Gonerby: Buddens Skylark Coaches, Woodfalls

This Bedford SB1, FBY 736, carries bodywork by Barbara and in this view is parked at the Sliema Ferries terminus waiting for its next departure on route 645 to Cirkewwa. Notice that the driver has removed the 6 from the route number box in the windscreen in the interim. "Id-Dielja" means the "the vine tree", and is probably a family nickname.

FBY 736	Bedford SB1	Barbara (1973)	B39F	7/60	3/73	AC Buses Ltd, Zabbar
FBY 737	Bedford YRQ	Duple Dominant I	BC45F	6/75	3/86	Joseph Vella, Mellieha
FBY 738	AEC Swift 4MP2R	Marshall	B46F	1/70	5/81	Joseph Refalo, Santa Venera
FBY 739	AEC Mercury (Maltese) f/c	Daina	B41F	4/67	4/67	Albert Sammut, San Gwann
FBY 740	King Long XQM 6113GMC	King Long	N45F	10/03	10/03	Renard Demanuele, Zabbar
FBY 741	Volvo B7RLE	Saracakis	N45F	5/03	5/03	Saviour Caruana, Qormi
FBY 742	Dennis Dart SLF	East Lancs	N45F	3/97	3/97	Leo Grech, Mosta
FBY 743	BMC Falcon	BMC	N45F	5/03	5/03	Jimmy Sammut, St Paul's Bay
FBY 744	AEC Swift 4MP2R1	Park Royal	B46F	10/70	6/81	Antoine Sant, Mellieha
FBY 745	Volvo B7RLE	Saracakis	N45F	6/03	6/03	Joseph Ellul, Mqabba
FBY 746	Bedford YRQ	Duple Dominant I	BC45F	5/73	by 11/84	Alfred Mercieca, Marsa
FBY 747	Bedford YRQ	Plaxton Panorama Elite II	BC44F	2/71	by 3/85	Francis Galea, Fgura
FBY 748	AEC Reliance 4MU3RA	Brincat (1976)	B45F	3/62	1/77	Charles Cortis, Naxxar
FBY 749	AEC Mercury 2GM4RA	Debono (1967)	B40F	1959	5/67	Silvio Buttigieg, Fgura
FBY 750	Dodge KR900	Aquilina (1972)	B43F	1966	1972	Joseph Gatt, Qormi
FBY 751	Bedford YRQ	Duple Dominant I	BC45F	2/75	by 4/85	Richard Borg, Mosta

FBY 736, Y-0736, Y-1437, A-3307, 3307 — Royal Navy, 97RN69, originally Strachans B36F bodywork
FBY 737, Y-0737, HGM822N — Reliance, Newbury, 147
FBY 738, Y-0738, Y-1439, A-3309, AML9H — London Transport, SM9
FBY 739, Y-0739, Y-0903, Y-1603, A-0434, 434 — Unscheduled bus, Malta (Y-0903). AEC chassis assembled in Malta
FBY 740 — New
FBY 741 — New
FBY 742 — New
FBY 743 — New
FBY 744, Y-0744, Y-1445, A-3315, EGN258J — London Transport, SMS258
FBY 745 — New
FBY 746, Y-0746, NMB279L — Romani, Bridgwater: Godfrey Abbott Group, Sale
FBY 747, Y-0747, ABW183J — Calloway, Rowley Regis: Carterton Coaches of Carterton
FBY 748, Y-0748, Y-0906, Y-1606, A-5574, 5574, 336VLG — Unscheduled bus, Malta (Y-0906): Derwent, Swalwell Bostock, Congleton
FBY 749, Y-0749, Y-1450, A-3320, 3320 — Magro, Fgura: AEC Mercury chassis was acquired in 10/64
FBY 750, Y-0750, Y-0904, Y-1604, A-5301, 5301 — Unscheduled bus, Malta (Y-0904): lorry chassis acquired new in 1966
FBY 751, Y-0751, Y-1452, GUX401N — Trefaldwyn, Montgomery: Corvedale, Ludlow, 46

Solaris FBY 752 negotiates the one-way system in Mosta on its approach to Mosta Dome on its way to Gharghur on route 56 at about 0930 on 2nd September 2009. AEC Swift FBY 675 follows behind en route to Bugibba and Qawra on service number 58.

FBY 752	Solaris Valletta	Solaris	N45F	5/03	5/03	Mario Mifsud, Mgarr
FBY 753	BMC Falcon	BMC	N45F	6/03	6/03	Alphonse Abela, Ghaxaq
						(Peppin Garage Ltd)
FBY 754	Bedford YRQ	Duple Dominant I	C45F	1/75	by 1/85	Horace Vella, Qormi
FBY 755	BMC Falcon	BMC	N45F	9/03	9/03	Joseph Cassar, Kalkara
FBY 756	AEC Swift 4MP2R1	Park Royal	B46F	10/70	7/81	David Borg, Mellieha
FBY 757	Bedford YLQ	Duple Dominant I	BC45F	2/77	by 5/92	Grezzju Borg, Zabbar
FBY 758	Volvo B7RLE	Saracakis	N45F	8/03	8/03	Raymond Schembri, Qormi
FBY 759	MAN 14.220	Scarnif	N45F	3/03	3/03	Francis Attard, Luqa
FBY 760	MCII	MCII	N45F	6/03	6/03	Sandro Abela, Zejtun
FBY 761	BMC Falcon	BMC	N45F	12/03	12/03	C.M.S. Transport Ltd
						(Stephen Cilia), Gharghur
FBY 762	Bedford YRQ	Plaxton Supreme III Express	C45F	3/76	by 11/84	Andrew Abela, Zejtun
FBY 763	AEC Swift 4MP2R	Marshall	B46F	2/70	6/81	Adrian Caruana, Zabbar
FBY 764	King Long XQM 6113GMC	King Long	N45F	11/03	11/03	Joseph Vella, Floriana
FBY 765	BMC Falcon	BMC	N45F	10/03	10/03	Carmelo Abela, Zejtun
FBY 766	Bedford SBG	Debono (1968)	B40F	1955	3/68	Charles Vella, Mosta
FBY 767	AEC Reliance MU3RV	Debono (1970)	BC46F	1955	12/70	John Camilleri, Naxxar
FBY 768	Dennis Falcon HC	East Lancs	BC43F	2/85	1/02	Albert Refalo, Attard
FBY 769	King Long XMQ6113GM (m)	King Long	N45F	5/03	5/03	Sebastian Ciappara, Qormi
FBY 770	Leyland Tiger TS	Barbara (1958)	B40F	by 12/39	7/58	Mario Cassar, Birzebbugia
FBY 771	BMC Falcon	BMC	N45F	7/03	7/03	Charles Carobott, Zejtun
FBY 772	AEC Reliance MU3RV	Aquilina (1971)	B40F	1955	3/71	Victor Meli, Mosta
FBY 773	AEC Swift 4MP2R	Marshall	B46F	5/70	7/81	Joseph Borg, Zabbar
FBY 774	AEC Mercury 2GM4RA	Aquilina (1965)	B40F	1958	10/65	Sebastian Mifsud, Mellieha
FBY 775	Dodge T110L f/c	Zammit	B40F	by 12/59	by 12/59	Filippa d'Amato, Zebbug
FBY 776	AEC Swift 4MP2R1	Park Royal	B46F	10/70	11/82	Paul & Rocco (Marsa Ltd), Marsa
FBY 777	AEC Mercury GMR4H	Falzon	B40F	1958	11/65	Charles Vella, Mosta
FBY 778	Bedford SB	Brincat	B40F	1954	10/72	Martin Borg, Mellieha
FBY 779	AEC Swift 4MP2R1	Park Royal	B46F	11/70	7/81	Patrick Cauchi, Mellieha
FBY 780	Bedford YRQ	Duple Dominant Express	BC45F	8/74	by 4/85	Albert Sammut, San Gwann
FBY 781	BMC Falcon	BMC	N45F	3/03	3/03	Etienne Falzon, Siggiewi
FBY 782	King Long XMQ6113GMC	King Long	N45F	7/01	9/01	Andrew Abela, Zejtun
FBY 783	BMC Falcon	BMC	N45F	6/03	6/03	Felix Fenech, Mosta

A quiet Sunday afternoon in September 2009 FBY 760, one of the small number of MCII low-floor vehicles on Malta, is seen heading past the Porte des Bombes to Kalkara on route 4. The car bearing left is heading for Msida, the route followed by many of the bus routes to Sliema, and those to Mosta, Qawra, Mellieha and Cirkewwa.

FBY 752	New
FBY 753	New
FBY 754, Y-0754, Y-1455, GUX400N	Wigston Coach Hire: M&M, Highley, 34
FBY 755	New
FBY 756, Y-0756, Y-1457, A-3330, EGN275J	London Transport, SMS275
FBY 757, Y-0757, YAA261R	Marchwood, Totton
FBY 758	New
FBY 759	New
FBY 760	New
FBY 761	New
FBY 762, Y-0762, Y-1463, LRG65P	Williams, Cwmdu: Rochester and Marshall, Great Whittington
FBY 763, Y-0763, Y-1464, A-3339, AML34H	London Transport, SM34
FBY 764	New
FBY 765	New
FBY 766, Y-0766, Y-1467, A-3342, 3342	Royal Navy, 39RN96 with B36F bodywork
FBY 767, Y-0767, Y-0871, Y-1571, A-2573, 2573, OWT940	Unscheduled bus, Malta (Y-0871): Anderton, Keighley
FBY 768, PIL 9738, B51XFV	Pilkington, Accrington: Hyndburn, 51
FBY 769	New
FBY 770, Y-0770, Y-1471, A-3347, 3347	Pre-WW2 Leyland bus chassis
FBY 771	New
FBY 772, Y-0772, Y-0873, Y-1573, A-2538, 2538, OOU 235	Zarb Coaches, Malta (Y-0873): Creamline, Bordon
FBY 773, Y-0773, Y-1474, A-3350, AML 50H	London Transport, SM50
FBY 774, Y-0774, Y-1475, A-3351, 3351	Second-hand AEC Mercury lorry chassis
FBY 775, Y-0775, Y-1476, A-3352, 3352	New as WW2 military Dodge n/c lorry chassis
FBY 776, Y-0776, Y-1477, EGN238J	London Transport, SMS238
FBY 777, Y-0777, Y-1478, A-3354, 3354	AEC Mercury chassis and cab imported from GB in 7/64
FBY 778, Y-0778, Y-0923, Y-1623, A-1788, 1788	Unscheduled bus, Malta (Y-0923): imported Bedford SB chassis
FBY 779, Y-0779, Y-1480, A-3356, EGN280J	London Transport, SMS280
FBY 780, Y-0780, OHB470N: (FBY 715 was carried during 2000)	Williams, Cwmdu, 23: Morlais, Merthyr Tydfil
FBY 781	New
FBY 782	New
FBY 783	New

FBY 791 is one of the small number of route buses in Paramount Garage's fleet. It operates only on Private services and is seen at Valletta Castle Gate terminus during a Sunday afternoon in September 2009.

FBY 784	Maltese f/c	Debono	B40F	1/60	1/60	Phyllis Caruana, Zebbug
FBY 785	Leyland Tiger Cub PSUC1/13	East Lancs (1967)	B45F	1/67	by 8/81	Charles Vella, Mosta
FBY 786	Bedford YLQ	Duple Dominant II	BC45F	6/77	1/88	Helen Buttigieg, Zabbar
FBY 787	Leyland Tiger Cub PSUC1/13	East Lancs (1967); Scarnif Broadway (2001)	B45F	2/67	by 8/81	Jason Borg, Zebbug
FBY 788	King Long XMQ6113GMC	King Long	N45F	6/03	6/03	Francis Bugeja, Zabbar
FBY 789	*This registration is currently not in use.*					Maria Antonia Zammit, Qormi
FBY 790	Bedford YRQ	Plaxton Supreme III	BC45F	2/76	7/85	Maria Antonia Zammit, Qormi
FBY 791	Dodge KEW	Aquilina (1975)	BC40F	by 12/66	9/75	Sean Grech, Naxxar
FBY 792	Leyland Lynx	Leyland	N45F	12/89	2/08	Z. S. Co. Ltd, Naxxar
FBY 793	Bedford YRQ	Duple Dominant I	BC45F	7/73	by 9/85	Carmelo Abela, Rabat
FBY 794	BMC Falcon	BMC	N45F	5/03	5/03	Angelo Sciberras, Qrendi
FBY 795	BMC Falcon	BMC	N45F	5/03	5/03	Joseph Cutajar, Mqabba
FBY 796	Bedford YRQ	Plaxton Panorama Elite III	C45F	9/74	by 11/84	Rennie Bonnici, Ghaxaq
FBY 797	AEC Swift 4MP2R1	Marshall	B46F	1/71	by 8/81	Z. S. Co. Ltd, Naxxar
FBY 798	AEC Reliance 2MU4RAE	Barbara	B46C	2/64	2/64	Antoinette Abela, Zejtun
FBY 799	*This registration is currently not in use*					Grezzju Borg, Zabbar
FBY 800	King Long XMQ6113GMC	King Long	N45F	11/03	11/03	Carmel Sant, Mellieha
FBY 801	King Long XMQ6113GMC	King Long	N45F	6/03	6/03	Isabelle Caruana, Zabbar
FBY 802	King Long XMQ6113GMC	King Long	N45F	6/04	2007	Joseph Borg, Zabbar
FBY 803	Bedford SLCO	Barbara (1973)	B39F	1960	7/74	Stefan Sammut, Msida
FBY 804	Bedford SB or SL	Debono (1967)	B40F	1952	4/67	Joseph Vella, Floriana
FBY 805	Optare Excel L1060	Optare	N45F	2/97	2/97	Paul and John Camilleri, Naxxar
FBY 806	King Long XMQ6113GMC	King Long	N45F	10/03	10/03	Paul & Rocco (Marsa Ltd), Marsa
FBY 807	Bedford YRQ	Marshall	B46F	7/75	by 4/89	Andrew Abela, Zejtun

Optare Excel FBY 805 is one of the original demonstrator vehicles which arrived on Malta in 1997. It is parked at Qawra Bus Station awaiting its next run to Valletta on the main route 49.

FBY 784, Y-0784, Y-1485, A-3495, 3495, 6512	Gozo post bus
FBY 785, Y-0785, Y-1486, A-3548, FCB12D	Blackburn, 12
FBY 786, Y-0786, SCD899R	Plumpton Coaches
FBY 787, Y-0787, Y-1488, A-3673, GBV18E	Blackburn, 18
FBY 788	New
FBY 789	
FBY 790, Y-0790, NFW 574P	Bird, North Hykeham: Clipson, Newton on Trent
FBY 791, Y-0791, Y-0852, Y-1552, A-3440, 3440	Paramount Garages, Mosta, (Y-0852): Dodge lorry chassis
FBY 792, G183EOG	Travel West Midlands, 1183
FBY 793, Y-0793, MUE648L	Woodcock, Buxton: Monty Moreton, Nuneaton
FBY 794	New
FBY 795	New
FBY 796, Y-0796, Y-1497, SJR428N	Dobson, Whickham: Curtis, Dudley
FBY 797, Y-0797, Y-1498, A-5293, EGN194J	London Transport, SMS194
FBY 798, Y-0798, Y-1499, A-5294, 5294	Chassis acquired new in 9/63
FBY 799	
FBY 800	New
FBY 801	New
FBY 802	New (This vehicle did not enter service until 2007)
FBY 803, Y-0803, Y-1504, A-5576, 5576	War Department, imported as a chassis in 5/66
FBY 804, Y-0804, Y-0862, Y-1562, A-2482, 2482	Unscheduled bus, Malta (Y-0862) on an imported chassis in 1/64
FBY 805, Y-0805	New
FBY 806	New
FBY 807, Y-0807, HGM616N	Atomic Weapons Research Establishment, Aldermaston

BUS ROUTES
from Valletta

KEY

68 68	Principal Valletta routes
12 12	Part-day or infrequent Valletta routes

Routes 132 and 153 are limited-stop 'Express'
Certain works / schools services are not shown.
Infrequent services only to be part shown.

© Brendan Fox 2010

OTHER ROUTES
not serving Valletta

SCHEDULED BUS ROUTES ON MALTA

Valletta dominates the bus scene on Malta. The provision of services from towns and villages across the island results in a hive of activity around the Triton Fountain throughout the day. Nowadays well over half the fleet of 508 route buses is rostered to provide each day's timetable. The great majority of the buses will therefore be plying the routes into the capital. Many of the remainder of the vehicles on a day's roster will be providing either the "Direct" routes which are based, principally, on the holiday resorts of Sliema, Bugibba and Qawra, or the comparatively recently introduced routes from villages to the Mater Dei Hospital and nearby Malta University. The remaining handful of route buses will be found on the few inter-town routes. All the routes currently timetabled are listed in the following pages, with details of places to see and visit en route.

The City Gate bus terminus just outside the main gate into Valletta is dominated by the Triton Fountain around which about half of the departure points are situated. The fountain is a comparatively recent addition to the area, being erected only in the late 1950s to a design by local artist Vincent Apap. Food and drink stalls are close by and intending passengers can obtain timetable information at a number of ATP kiosks. Route Buses and Unscheduled Buses (tour coaches) lay over in the extensive parking area overlooked by St James's Bastion. Following the recent introduction of the tourist routes, the open-top double-deckers of Cancu Supreme and Garden of Eden can be seen passing through too. Visitors will also find the Park and Ride terminus in the vicinity, as it is only a short walk from Castle Gate along Triq ir-Repubblika close to the ruins of the old Opera House.

The historical and tourist sights of the capital deserve a Handbook all their own. However, it would be remiss not to recommend St John's Co-Cathedral and Museum with its paintings by Caravaggio, the Grand Master's Palace, Palace Armoury, Upper and Lower Barracca Gardens and the old Saluting Battery (one of the main defences of Valletta built in 1500s) where the Noonday Gun is fired every day.

Parked by the Triton Fountain, DBY 433, one of the five Solaris Valletta low-floor vehicles in the Route Bus fleet, awaits its next departure from Castle Gate bus terminus. This was the vehicle supplied with the incorrect EBY 433 registration when originally placed in service in late 2002.

There were, originally, eight Auberges of the Knights of St John in Vittoriosa, before the Knights moved to grander premises in Valletta when the capital was built, each one representing one of the eight languages (langues) of the Order of St John. These auberges (hostels) may be compared with present-day Oxbridge colleges and they included a chapel, dining-hall and accommodation built around an inner courtyard. Gerolamo Cassar designed them all, and nowadays two are open to the public. The Auberge d'Italie is now the General Post Office in the capital.

The Bastions built along the ramparts surrounding Valletta form a defensive system with curtains and forts, too. The Wartime Experience is an audio-visual show about the Siege of Malta in 1942, and the Manoel Theatre, reputed to be the third oldest in Europe, and its adjoining museum cannot be missed. The views over the stunning Grand Harbour cannot fail to impress and bus travellers to the Three Cities will have numerous opportunities to see the capital from the opposite shores of the Harbour.

All the route buses set off from Castle Gate along Vjal Nelson and after negotiating the War Memorial roundabout, they enter Floriana which takes its name from the Italian engineer, Paolo Floriani, who designed the fortifications which were built as Valletta expanded during the seventeenth century. These fortifications were needed in the light of threats of invasions by the Ottoman Turks and were constructed during the fifteen years from 1635 to 1650. Along the main dual carriageway of Triq Sant Anna are the US Embassy and British High Commission buildings. The Porte des Bombes stands at the western end of Floriana. It is an elegant ceremonial Baroque gateway with three arched doorways and was constructed between 1697 and 1720, thus forming part of the city's outer defences. Nowadays it is an imposing traffic island with the curtain walls cut away for traffic to pass on either side.

At this point, it would be remiss not to explain the route taken by buses approaching Valletta. After passing the Porte des Bombes (buses are banned from passing through the arches themselves), vehicles climb up towards Triq Sant'Anna, but then turn left so that they do not travel along Triq Sant'Anna on their approach to the terminus. Instead they travel along Triq Sarria, passing the Argotti Botanical Gardens with their rare species of cacti and the Maglio Gardens, a tree-lined walk alongside the road. To the right of Triq Sarria is a huge open area, larger than a football field, called the Granaries. This is where the Knights of Malta stored a two-year supply of grain for the island in underground chambers, the stone lids of which are still clearly visible. Nowadays, the Granaries are an open area large enough to hold a crowd of up to 100,000, especially for political demonstrations and, in May 2001, for the visit of Pope John Paul II.

Resuming the outward journey immediately after the Porte des Bombes, route buses start to fan out in three directions towards their ultimate destinations. The first to divert are buses on routes via Msida to Birkirkara, Mosta, Golden Bay, Bugibba, Mellieha and Sliema. These buses swing left passing under the main road. On the downhill run towards Pieta, buses pass Ta' Braxxia Anglican Cemetery and then skirt the quays at Pieta and Msida along a tree-lined road with a sharp left-hand bend on the approach to Msida. At the huge roundabout at Msida, buses turn right for Sliema. Most other buses (basically the routes to Mosta and beyond) continue straight ahead. After passing beneath the Regional Road, buses climb up Triq il-Wied ta'l-Imsida to Birkirkara, which is the largest township on the island with a population of over 23,000 living in its narrow, crowded streets.

Meanwhile, back at the Porte des Bombes, all other buses stay on the dual carriageway for a few hundred metres until they pass the Park and Ride car-parks at Blata l-Bajda. Buses on routes numbered from 1 to 39 continue along Triq Nazzjonali heading for Marsa and Paola, whilst routes from 70 to 89 (basically) head for Hamrun and beyond, by crossing the fly-over over Triq Nazzjonali.

ROUTES 1, 3, 4 and 6 to THE THREE CITIES and KALKARA

Vittoriosa, Cospicua and Senglea are situated in the Cottonera district, lying south of the Grand Harbour to the east of Valletta. They were granted the all-embracing name of the Three Cities by Napoleon's commander, General Vaubois, in 1798, in the vain hope that, by embellishing the residents' status, he would be able to win them over to accept new French legislation which was being introduced to the island.

The Three Cities had originally been settled by the Order of the Knights of St John after 1530; the Vittoriosa area of the Grand Harbour was known, in Italian, as "il Borgo" (the village), which the Maltese soon altered to a local variant "Birgu". Senglea is known as L'Isla, and Cospicua as Bormla. During the next century and a half, the Order regarded the creeks as strategically important in the defence of the island against Turkish aggression. The Margherita Lines and Cottonera Lines which are a series of eight bastions around the landward perimeter of Senglea, Cospicua and Vittoriosa were built during the mid-seventeenth century.

During the Second World War, German and Italian aircraft bombed the dockyards situated on the creeks where these three towns lie. In April 1942, more than 3,100 tons of bombs fell on the Cottonera area; consequently little of historical value remains nowadays. The area was quickly rebuilt after the war, to enable the residents who had been evacuated to return home.

Cospicua links Vittoriosa with Senglea and its ring of forts protect the Cities on the landward side. Originally named Bormla, it was renamed after its "conspicuous" part in the Great Siege.

Senglea is named after the Grand Master Claude de la Sengle who allowed families to settle on the newly-fortified peninsula during the 1550s, if they then built their new homes on the building plots which the Grand Master had allocated to them free of charge. The Turks mounted a huge attack on the peninsula about ten years later, in 1565. Over 8,000 of them were killed, but Fort St Michael was strong enough to withstand their attacks. Bombed during World War Two, the peninsula has since been rebuilt, as has the Church of Our Lady of Victories which had been badly damaged during an air-raid in 1941 which targeted the aircraft carrier HMS Illustrious undergoing repair in Dockyard Creek. The Dockyards are now mostly closed down, so much so that the long-established dockyard bus services have been withdrawn in recent years. The walk to Senglea Point on the tip of the peninsula is well worth making. Here, the "vedette", carved with ears and eyes symbolising watchfulness, commands a splendid view over the Grand Harbour and parts of Valletta.

Vittoriosa was the home of the Knights from 1530 until about 1571. The Inquisitor's Palace was originally the law courts and became the residence of the Inquisitor, the Pope's Apostolic representative, whose appointment on Malta was to oversee the enrolment of suitable Christians into the Order of St John. The Palace reflects the lives of these Inquisitors. The Collegiate Church of St Lawrence was where the lifting of the Great Siege of 1565 was celebrated; its relics include a silver processional cross from Rhodes which was used in the Crusades, and is always paraded during the City's *festa* on August 10th each year. Fort St Angelo is built on the original site of a Phoenician temple. During the Great Siege it was the focal point of the Turks' attacks. Later it became a prison, but its principal outline, following additional defences being built in the late seventeenth century, remains today. Under British rule the Fort was, successively, the HQ of the commandant of the barracks, a naval barracks, naval station and submarine base.

EBY 528 is seen climbing steadily up the steep hill towards the Cospicua terminus mentioned in the text. It is a Maltese/Bedford vehicle with Debono bodywork. The bus would however continue to Kalkara on route 4. The minor basilica of the Nativity of the Blessed Virgin in Senglea dominates the skyline of this picture which shows a typical scene in the immediate environs of the Three Cities.

Kalkara lies outside the bastions which protect the Three Cities, on the peninsula between Kalkara Creek and Rinella Creek. It has a small harbour where traditional Maltese boats are repaired or moored during the winter.

After the Porte des Bombes buses soon reach Triq Aldo Moro (a four-lane stretch of road) with the dockyards on the left. Marsa is also the site of Malta's only racecourse, where the island's trotting races are held. Trotting is one of the island's top spectator sports during the season which lasts from October to May. Arab horses compete in these races, and they are often to be seen on the public roads with their owners precariously perched on their trotting-carts hurtling along at break-neck speed.

Buses soon begin the climb up to Paola, past the island's prison and nearby mosque. Paola was established in 1626 by the Grand Master Antoine de Paule to cope with overspill from Valletta and the Three Cities. Buses turn left along Triq Bormla where the balconies of the houses lining both sides of the road are worth admiring. The main road past the Dockyard entrance soon passes through the tunnel under St Paul's Bastion. Route 4 continues down the hill to Cospicua Square, whereas routes 1 and 6 turn right up the steep gradient of Triq San Gwann T'Ghuxa. Route 6 then passes through St Helen's Gate on the approach to Cospicua, as do all buses returning from Cospicua Square to Valletta, because of the one-way system. Travelling alongside Cottonera Marina buses soon climb, bearing right, to the terminus opposite the Couvre Porte (one of main entrances to Vittoriosa town). Don't be misled if you hear this terminus being referred to as Cospicua and Vittoriosa — it is the same place!

Buses on route 1 serve the Verdala area of Cospicua by travelling the whole length of Triq Alessandra. The departures from Valletta at xx15 also serve the Tal Hawli estate, thus giving easy access to De La Salle College from where a number of contract coaches and route buses can be seen departing at the end of the schoolday. The complicated timetables for this area show that Tal Hawli is served on return journeys to Valletta only until 1000.

EBY 593 is a Bedford YRQ with Duple Dominant I bodywork which was new to M&M in Highley near Bridgnorth in Shropshire. It is seen operating a schools service from De La Salle College which is situated near the Three Cities, in June 2009. *T W W Knowles*

Route 3 to Senglea has two distinct variations; the more direct route follows routes 1, 4 and 6 through the tunnel under St Paul's Bastion, but then continues straight ahead down the hill in the one-way system along Triq San Frangisk, before turning left through St Michael's Bastion, and then sweeping down to the terminus along Triq il Vittoria. The other longer route to Senglea (both variants are labelled route 3 and so are indistinguishable to the casual passenger) serves Fgura's main street, Triq iz-Zabbar, and Triq San Tumas. Thanks to a southbound bus-lane, buses on this route are able to serve this one-way street in both directions. The approach to Cospicua is under the Polverista Gate, after which the buses turn left down Triq San Gwann T'Ghuxa, and then turn right to rejoin the "direct" version of route 3, en route to Senglea. The return journey to Valletta on route 3 passes under St Helen's Gate and back down Triq San Gwann T'Ghuxa, because of the one-way system in the area.

Route 1 Valletta – Marsa – Paola – Verdala – {Tal Hawli} — Vittoriosa

From Valletta, departures (basically) every 30 minutes between 0720 and 2120 Monday to Saturday, and between 0715 and 2130 on Sunday. Journeys departing Valletta at xx15 or xx20 serve Tal Hawli.

Route 3 Valletta – Marsa – Paola – {Fgura} — Cospicua — Senglea

Departures are from 0600 (0700 on Sundays) to 2200 from Valletta, and from 0530 (0600 on Sundays) to 2130 from Senglea, every 20 or 30 minutes. Fgura journeys are hourly between 0600 and 2100 (0700 to 2100 on Sundays). From Senglea all timings serve Fgura until 1000, then hourly until 2100.

Route 4 Valletta – Marsa – Paola – Cospicua Square — Vittoriosa — {Rinella} — Kalkara

On weekdays departures are between 0700 and 2100 from Valletta, every half-hour, and at weekends between 0700 and 2130. Throughout the week departures between 0930 and 1630 serve Rinella.

Route 6 Valletta – Marsa – Paola – Cospicua Square — Vittoriosa

Departures from Valletta from 0530 until 2200 roughly each hour, but with extra early morning timings.

On route 4, the bus follows the direct version of route 3 as far as the left turn to Senglea, before travelling through Cospicua's main square and then climbing up to the terminus of routes 1 and 6. Continuing straight ahead, it soon approaches the archway into Kalkara and makes its way along Triq San Dwardu and turns left into Triq Santa Liberata so as to overlook Kalkara itself down at sea-level. Every half-hour vehicles turn right after the Cappuchin Convent to serve the entrance of the Mediterranean Film Studios with its 10m deep-water tank built for the filming of Raise the Titanic, Fort Rinella or Battery with its huge 100-ton Armstrong gun, and Rinella Creek. This route variation is now operated every day of the week, having originally been Mondays to Saturdays. These Rinella variations eventually rejoin the normal route 4 which now travels past the former Royal Navy Hospital Bighi which is now a school, before descending Triq Marina to travel alongside Kalkara Creek, past the grand St Joseph's Church. From here there is a short climb to a picturesque arch back into Birgu and the terminus of routes 1 and 6. Passengers travelling back to Valletta sometimes have to change buses here. Routes 1, 4 and 6 combine to provide a bus every fifteen minutes (and every ten at peak times) between Valletta and Vittoriosa for most of the day. On the inward journey to Valletta, all vehicles on this group of routes serve Paola Square itself.

Route 5 Valletta — Marsa — Kordin Industrial Estate

One works journey M-F at 0630.

Route 14

On Saturday and Sunday evenings three departures from Valletta serve Verdala, Tal Hawli and the Kalkara loop, hence the routes 1 and 4 being combined to give this route number. Timings are at 2040 SSu and 2130 SuO.

Route 16 Valletta — Marsa — Paola Secondary Schools

Schooldays only service at 0710 and 0745.

DBY 352 has Welsh origins, having been with Davies Bros. of Pencader when new. In September 2009, it is seen leaving Senglea, having just passed through the imposing gateway on its return journey to the capital on route 3. Flags had been erected for the *festa* taking place the following Tuesday.

ROUTES to LUQA (AIRPORT) and BIRZEBUGGIA

Route 8 Valletta – Marsa – Paola — Tarxien – Bir id-Deheb – Ghaxaq – Gudja — Luqa (Malta Airport) (circular).

Buses on this group of routes also follow the same route from Valletta to Paola as routes 1 to 6 above. Soon after passing the prison, buses turn right into Paola Square, passing in front of the Church of Christ the King, a huge mock-Romanesque building of the 1920s. Paola was founded in the seventeenth century as an overflow town for the population of the Three Cities. But, because it was built outside the fortifications, the town developed only slowly.

Paola is famous for the Hal Saflieni Hypogeum, an underground burial site, which is considered to be the island's finest archaeological monument. Discovered in 1902 during housing development, when builders who were cutting cisterns broke through its roof, the Hypogeum was the burial place for about 7,000 bodies, along with personal ornaments and pottery. The "Holy of Holies" chamber is thought to have been a burial place and a shrine. This temple complex is believed to have been in use for seven centuries from 3,700 BC until 3,000 BC. Visitors to the Hypogeum are strictly limited to seventy a day, and booking (up to a month ahead of a visit) is strongly recommended.

On leaving Paola, the bus soon reaches Tarxien which is famous for its Neolithic Temples dating from about 3,100 BC, and which were excavated between 1914 and 1919. Even the town's name reflects its history, as "tirxa" in Maltese means "stone slab". In the South Temple visitors will find the statue of a giant goddess — the most famous relic in the museum – a large woman whose upper torso is missing, but whose full pleated skirt covers broad hips and legs. Tarxien is famed for its altars, sacrificial blocks, blocks decorated with animal friezes, stone bowls for burning aromatic herbs, beautifully paved courtyards and solidly carved wall slabs.

After negotiating Tarxien's narrow streets and one-way system, the bus eventually reaches Triq Tal Barrani, where it turns left and continues straight on until the roundabout at Tal-imhammed near the premises of the Garden of Eden coach firm. Turning right here soon reaching the narrow roads through Gudja. The bus joins the main road skirting Luqa Airport, and then threads its way through the Airport grounds to the terminus in front of the main Terminal Buildings. Gudja is served only on outbound timings from Valletta. Buses in Ghaxaq skirt the village centre by taking the bypass, although a recent short-lived deviation saw most inbound buses serve the village centre for a time.

Gudja claims to be the birthplace of Gerolamo Cassar (1520 – 1586) who was the architect of many of Valletta's finest churches. In Main Street stands Palazzo d'Aurel, one of the island's finest country palaces, built during the eighteenth century. It is reputed that Napoleon Bonaparte and Horatio Nelson, accompanied by Sir William and Lady Hamilton were entertained here. On the outskirts of the village is the small Church of St Mary Ta'Bir Miftuh, one of the oldest remaining churches on the island, dating from 1436.

Departures from Valletta are every 20 or 30 minutes from 0630 until 2100, (0600 until 2130 at weekends) and from 0550 until 2000 (0530 until 2030 at weekends) from the Airport. Later journeys from the Airport are provided by route 34.

Route 11 Valletta – Marsa – Paola Square – Tarxien – Bir id-Deheb – Birzebbugia – Pretty Bay

This route follows exactly the same roads as route 8 as far as the roundabout near the Garden of Eden premises. Instead of turning right at the roundabout, the bus then continues straight ahead past Ghar Dalam caves (q.v. route 127 to Marsaxlokk) to Birzebbugia. Once a fishing port, the town developed as a summer resort for the islanders

DBY 324 sports a Casha body on a Ford V8 lorry chassis. Here it is operating route 12 in the Tal Papa housing estate west of the Pretty Bay terminus.

long before the nearby Malta Freeport was developed. Buses turn left to follow the promenade coast-road round to Pretty Bay. This bay boasts a broad expanse of cream-coloured sand alongside the bus terminus. The sand is always "topped up" by the local council after the winter. Palm trees line the roads in the town. The Pretty Bay name is something of a misnomer, as the Freeport is 800 metres distant across the water and so cranes and the container port dominate the nearby coastline.

Departures from Valletta begin at 0600 and finish at 2230; the first bus leaves Birzebbugia at 0530 and the last at 2130. Timings, combined with other routes to and from the town (routes 10, 12, 13, 113 and 115), throughout the day are every 15 or 20 minutes. Return journeys omit Paola Square, and so travel along Triq il Palma instead.

Thus, there is also:

Route 10 Birzebbugia — Bir id-Deheb — Tarxien — Paola Square — Marsa — Valletta

This is an inward variation of route 11 via Paola Square weekdays only, approximately every hour.

Route 12 Valletta – Marsa – Paola Square – Tarxien – Bir id-Deheb – Birzebbugia — Tal Papa Housing Estate (returning to Pretty Bay terminus). Thus, this is an outbound only service, apart from one early morning journey.

Route 12 operates to Pretty Bay at times which complement route 11. From the Pretty Bay terminus, the bus travels along Triq San Patrizu and turns right and then left to make a clockwise circuit of the large residential area of Tal Papa. Departure times from Valletta are M-F at 0645, 1130, 1340, 1530 and 1950. Saturday timings are 1130, 1240, 1345, 1530, 1730 and 1940, whilst those on Sundays are 1140, 1240, 1530, 1730 and 1945.

Routes 13 or 113 Valletta — Marsa — Paola — Tarxien — Bir id-Deheb — Birzebbugia — Hal Far

From the Pretty Bay terminus, route 13 proceeds past the Freeport out into a stretch of countryside before arriving at Hal Far Industrial Estate which is built on the site of the World War Two RAF airfield. After travelling along Triq Ghar Hasan through the Estate, the bus reaches the outer terminus at Hal Far AFM Headquarters and returns to Birzebbugia around a complex of wide roads through the Industrial Estate, passing the Playmobil FunPark for young children and the storage facility of the slowly developing Maltese Historic Vehicle Trust.

The timings on this route have been significantly enhanced since the first edition of this *Bus Handbook*. The terminus at Hal Far is nowadays noted for the large tented village for African immigrants who have arrived on the island following Malta joining the EU in 2004. Return journeys to Valletta omit Paola Square, too.

Route 113 (originally operating only on weekdays, but now daily) mirrors route 13 exactly on the outward journey, but returns direct to the Pretty Bay terminus, omitting the Industrial Estate complex.

Route 15 Valletta – Marsa – Paola Square – Santa Lucija (circular)

Again following route 8 as far as Paola Square, the bus then continues straight on along Triq Hal Luqa to a large roundabout on the outskirts of Tarxien. Immediately after the roundabout the bus turns sharp left into Santa Lucija, where the anti-clockwise loop through the village is regularly signposted by prominent arrowed "Bus Route" signs. The large and well-kept houses in this village were homes for members of the British forces during World War Two.

Departure times from Valletta are from 0620 until 2000 and from St Lucija from 0600 until the last return journey at 1920. An interesting feature of route 15's daily timetable is that there are no departures between 1300 and 1430 and after 1930. Instead route 115 operates, providing the alternative longer variation out to Pretty Bay.

Route 115 Valletta – Marsa – Paola – Tarxien – Santa Lucija – Tal-Barrani – Bir id-Deheb – Birzebbugia – Pretty Bay

This route covers the same roads to Santa Lucija as route 15 and then travels along the main Triq Tal-Barrani to rejoin route 11 out to Birzebbugia and Pretty Bay. It operates from Valletta between 1300 and 1430 and after 1930.

GOZO AIRPORT SHUTTLE

This route was introduced during the summer of 2007 and was operated by the ATP on behalf of Gozo Channel, the ferry operator. Low-floor buses were always allocated to the route which operated non-stop between Malta International Airport and Cirkewwa with drivers often taking the route through Manikata and along the Mellieha bypass. The route was withdrawn on Saturday, March 28th 2009. An on-line bookable minibus service, MaltaTransfer.com is now offered.

ROUTES TO ZABBAR AND MARSASCALA

The various routes to the Zabbar and Marsascala areas of the island all follow the roads covered by route 3 as far as Fgura. At the roundabout at the junction of Triq iz-Zabbar and Triq Hompesch where route 3 turns left, all these routes carry straight on to the Hompesch Arch roundabout on the outskirts of Zabbar. This imposing archway was the last monument built by the Knights of St John and it commemorates the day when Grand Master Ferdinand von Hompesch consented to the proposal to raise the village's status to a town.

Vehicles continue straight ahead, swinging left along Triq Tal-Labour, then turn right into Zabbar town centre with its narrow one-way system. The splendidly ornate church, the Sanctuary of Our Lady of Graces, which was designed by Tommaso Dingli, dominates the town centre. Its first stone was laid on 10th May 1641 and the church was completed fifty-five years later, with another fifty-nine years passing by before the gilding and ornamentation were finished. Its original dome was badly damaged by the French in 1800, and the church also suffered badly during World War Two. It has become a place of pilgrimage for those asking for intercession from Our Lady, especially on the eve of the town's *Festa*.

Numerically the last Route Bus in the fleet, FBY 807 is a Bedford YRQ with Marshall bodywork. It is working route 18 to Zabbar and is seen passing the island's prison on the approach to Paola.

FBY 680 is another venerable bus now over fifty-five years old. It boasts a Dodge f/c chassis with Sammut bodywork and is seen operating route 19 at the bus terminus in Marsascala, as passengers alight during an afternoon in September.

The town's great miracle has an even more recent history. On 14th October 1975, an RAF Vulcan bomber exploded over the centre of the town. All three crew members were killed, but only one resident died, struck by an electric cable severed in the blast. To this day the residents regard the event as a miracle.

Marsascala was the favoured fishing port of the Sicilian community on the island — hence its name meaning "Sicilian harbour". It remains a quiet holiday resort which has maintained its identity as a fishing port, too. Triq ix-Xatt promenade is the town's principal restaurant and café area. On the north side of the bay is the Church of St Anne with its distinctive Italianate campanile.

Zonqor Point, which houses the national swimming pool where national water-polo matches are held, is close by, too.

St Thomas Bay is separated from Marsascala by the headland of Il Gzira, on which Fort St Thomas was built in 1614 at the huge cost of 13,450 scudi, following the last Turkish invasion onto Maltese soil, as a revenge attack on the Knights for plundering grain. The fort boasted four corner towers and a seaward battery garrisoned by one hundred men in time of siege. The ravages of time and the sea ate away the battery which was eventually demolished in the 1970s. The Jerma Palace Hotel was built on the site, but since the publication of the first edition of this *Bus Handbook*, this hotel has been shut and now stands derelict and dilapidated on the shoreline.

The seventeenth century Mamo Tower is situated on the main road between Zejtun and St Thomas Bay. It is a miniature fort built in the shape of a St Andrew's Cross.

Route 17 Valletta – Marsa – Paola Square – Fgura – Zabbar – Zonqor Point – Marsascala

This is an infrequent route which serves the roads on the northern side of Marsascala Bay. The route is very similar to that followed by route 19 (see below) from Valletta all the way to Zabbar and beyond. As route 19 does, it too leaves Zabbar along Triq haz-Zabbar,

FBY 786 is a Bedford YLQ with Duple Dominant II bodywork. It is seen parked for a few minutes along the sea-front at Xghajra, the terminus of route 21.

but bears left into Triq iz-Zonqor. The bus travels clockwise around this headland, before returning to Marsascala where route 19 is rejoined. The bus then travels along Triq is-Salini on the southern side of the Bay past the derelict hotel and returns to the Marsascala terminus via Triq il-Qaliet. Departures from Valletta on weekdays are at 0730, 0830, 0910, 1010 and then hourly between 1310 and 1910. At weekends timings from Valletta are 0715 (Saturdays only), 0830 and then approximately hourly from 0910 until 1910.

Route 18 Valletta – Marsa – Paola – Fgura – Zabbar

Route 18 operates from Valletta from 0600 until 2200 and from Zabbar from 0530 until 2200. Services to the locality, when combined with routes 19 and 20, are every twenty or thirty minutes.

Route 19 Valletta – Marsa – Paola – Fgura – Zabbar – Marsascala

Route 19 operates every 15 to 20 minutes from 0630 until 2100 from Valletta on weekdays, with later journeys at weekends. From Zabbar, the bus travels direct along the main Triq haz-Zabbar to Marsascala and then performs the loop via St Thomas Bay and Triq tal-Gardiel back to the terminus.

Route 20 Valletta – Marsa – Fgura – Zabbar bypass – Marsascala

Buses on this route approach Marsascala along the bypass (Triq Sant' Antnin) and then cover the loop via St Thomas Bay. This route leaves Valletta hourly from 0850 until 1950 on weekdays, and on the hour, from 0800 until 2000 at weekends.

DBY 322, a Fordson with Tonna bodywork of 1964 vintage, poses for a seaside shot, as it makes the circuit of the residential areas from Marsascala and St Thomas Bay one sunny afternoon.

Route 21 Valletta — Marsa – Paola Square – Fgura – Zabbar – Xghajra

Leaving Zabbar the bus heads along Triq ix-Xghajra and turns left to serve the residential area along Triq L-Ghakrux Marzebb. Soon after the roundabout at Ta' Maggi, the bus turns left towards Xghajra, winds through a residential area before reaching the seafront, along Dawret ix-Xatt, where it might halt for a few moments before returning direct to Zabbar and Valletta.

On weekdays, this service operates a circular service approximately every half-hour from Valletta from 0600 until 2020. The weekend service is only hourly.

Route 22 Cospicua - Marsascala

This is an unusual route, one which does not serve Valletta, yet is not classified as a "Direct Route" so as to attract higher fares.

The "Cospicua" terminus is in fact the Senglea terminus of route 3. The route serves Cospicua Square, passes the terminus in Vittoriosa of routes 1 and 6 and then De La Salle College, before passing through Notre Dame Gate, an ornate and historic gate built in 1675. The bus soon turns left, thus joining the main route to Zabbar from Valletta, and negotiating the one-way system through Zabbar and out to Marsascala, followed by a circuit of the roads out to St Thomas Bay, returning to Marsascala terminus via Triq Tal Gardiel.

The route operates hourly from Senglea between 0715 and 2115. Certain timings serve Zonqor (as does route 17), but the timetable adhered to by drivers during the summer of 2009 differed greatly from those posted at termini.

ROUTES serving ZEJTUN and MARSAXLOKK

Route 27 Valletta – Marsa — Paola Square — Tarxien – Zejtun – Marsaxlokk

The first section of this route covers the same roads as route 8 (q.v.). Once past Tarxien the bus regains the main road and after passing along Triq Tal-Barrani, the vehicle turns left to serve Zejtun.

Zejtun is an old agricultural village, known to have had its own parish as early as 1436. In May 1565 the first attacks of the Great Siege on the island took place here, when two young Knights were put to death by the invading Ottoman Turks, but not before they had given misinformation to their killers about the strength of the defending forces. Consequently the invaders made a number of strategic errors in their war plans. In 1614, well after the Great Siege, the Turks once again pillaged the town but were successfully and quickly forced back to their ships in Marsascala harbour by the Knights' cavalry. It was the last attack on the island by the Turks.

Lorenzo Gafa, the notable local Baroque architect, started the parish church of St Catherine in 1692. With its striking octagonal dome, the church is a fine example of his work, with Doric and Ionic pilasters and two bell-towers adorned with Corinthian pilasters. The older parish church of San Girgor dates from 1436 and boasts the oldest dome on the island. The main door to the church is built off-centre, so that the Devil — who walks only in straight lines – will not be able to enter to disturb a church service. Within the church's thick walls, two narrow passages were discovered, in 1969, where the bones of more than eighty inhabitants of the village lay. The villagers are believed to have hidden there from Turkish invaders in 1547 and died in their hideout, when the church was set alight.

After serving the village centre, past the convent and church of San Girgor, the bus passes the ribbon development of houses and offices at Bir id-Deheb close to the Garden of Eden premises. At the nearby traffic-lights, the bus turns left, and soon passes the vineyards of Marsovin, the island's principal vine-growers. Soon Marsaxlokk is reached. Passengers alight at the sea-front, close to the open-air market stalls which sell fish, vegetables, fruit, local lace and tablecloths, souvenirs and handicrafts daily. Sundays are the best time to visit the markets here; extra bus services are provided on the routes to Marsaxlokk to cater for the influx of tourists. Recent pedestrianisation of the Town Square has meant changes in the terminus of the routes in the village. Buses now lay over in Triq L'Arznell which runs parallel to the main esplanade. Inbound journeys return to Zejtun but then do not travel along the main road to Tarxien, instead heading along Triq Bormla via the Bulebel Industrial area on the southern outskirts of Fgura, before reaching Tarxien and the "regular" route back to the capital.

Marsaxlokk is a very pretty fishing village, where the traditional colourful *luzzu* fishing boats with a Phoenician Eye (a good luck charm) painted on the prow are always to be seen moored close inshore. Marsaxlokk was the site of the first attack by Mustapha Pasha's army of what was to become known as the Great Siege, on 19th May 1565. The 1614 invasion of the island was started here, too, but the Turkish invaders were frightened off by the guns of Fort St Lucijan and landed at nearby Marsascala, then to attack Zejtun.

In 1798 Marsaxlokk was one of the five landing-places chosen by Napoleon Bonaparte in his attempts to capture the island from the Order of St John. His forces landed without resistance. Later Lord Nelson and his fleet took on supplies here whilst blockading the French forces in Valletta.

Much more recently the harbour was the venue for the summit meeting between Mikhail Gorbachev and US President George Bush in 1989; a meeting which is now viewed

DBY 376 sports bodywork by a builder, Mifsud, whose craftsmanship is rarely seen on Maltese buses nowadays. This Bedford SB is now at least forty years old and is seen along the sea-front in Marsaxlokk on the route 27, the main service from Valletta.

as ending the Cold War. Locals still talk about the bad weather during the weekend of the summit, when the infamous winter gregale, a violent north-easterly wind, whipped up the waters in the harbour, where their naval ship was anchored.

Close to Marsaxlokk harbour stark examples of the island's recent industrialisation are clearly visible—a large electricity generating station near Delimara Point and the Freeport across Marsaxlokk Bay towards Birzebbugia.

The route operates every half an hour from 0630 until 1930 from Valletta, then approximately hourly until 2110 on weekdays and until 2130 at the weekend. However, on Sundays, there is a more frequent timetable to enable passengers to visit the large market on the seafront, with departures every ten to fifteen minutes between 0900 and 1600.

Route 127 Valletta – Marsa – Paola Square – Tarxien – Ghar Dalam Caves — Marsaxlokk

This service operates hourly from Valletta between 0730 and 1930. It differs from route 27 only in its approach to Marsaxlokk. Instead of turning left at the traffic lights near to the Garden of Eden premises, the bus continues straight on towards Birzebbugia. It soon passes the Ghar Dalam cave complex, where remains of dwarf elephants and hippopotamus dating back 180,000 years, bones and antlers of deer, and human remains from the Neolithic and Bronze Ages have been found.

Shortly before reaching the outskirts of Birzebbugia, the bus turns left through the residential area of Qajjenza before reaching more open land where St Lucijan Tower, erected in 1610, overlooks St George's Bay. This coast-road bears left towards the southern end of Marsaxlokk Bay and the bus then threads its way round the back of the town to the terminus in Triq l'Arznell.

King Long EBY 509 turns round at the terminus of route 30 to St Thomas Bay. This irregular route no longer serves the bay itself, so passengers for the seaside nowadays have a five-minute walk down to the coast. This lunchtime shot sees the vehicle about to return light to Zejtun.

Route 29 Valletta – Marsa — Paola – Tarxien – Bulebel — Zejtun

This route terminates in Zejtun and covers route 27's inbound journey in both directions. It operates from Valletta from 0600 until 2200 (2230 at weekends) and from Zejtun from 0530 until 2100 (2130 at weekends) . After reaching the terminus in Zejtun the bus performs a loop of the suburbs, passing close to the premises of Cancu Supreme, before returning to the terminus.

Route 30 Valletta – Zejtun – St Thomas Bay (circular)

This route, too, covers the same roads to Zejtun, from where it passes through the village centre and an extensive residential area of the town (as route 29), before turning left onto the country road leading to St Thomas Bay. Recent modifications to the route mean that the buses no longer travel to the bay itself but turn round at the road junction a few hundred metres distant. Buses then return direct to Zabbar, often out of service.

There are only six weekday departures from Valletta at 0715, 0815, 1015, 1220, 1415 and 1615; on Saturdays at 0715, 0815, 1215, 1415 and 1615; on Sundays at 0815, 1220 and 1420.

Route 427 Qawra – Mosta – Attard – Marsa – Paola – Tarxien — Marsaxlokk

This is one of the island's "Direct" services linking the holiday resorts on the north coast directly with Marsaxlokk without going via Valletta. From Qawra Bus Station the route passes along Bugibba promenade to St Paul's Bay, from where the main road to Mosta is followed (see Route 44). Continuing towards Valletta the bus then turns right at the roundabout close to Mosta Technopark skirting Lija before joining the main road through Attard served by route 40 and passing close to San Anton Gardens. The bus then turns onto a fast stretch of road along the Mriehel Bypass soon reaching Marsa with its

Volvo Scarnif FBY 727 featured in-build in the first edition of this *Bus Handbook* with just part of its front and registration plate peering out of the Scarnif factory. Here it is in its full splendour at the Marsaxlokk terminus of route 127 from Valletta.

Racecourse where route 27 is joined for the remainder of the journey to Marsaxlokk via Paola and Tarxien.

This route has only three timings from Qawra, at 0915, 1015 and 1600, returning from Marsaxlokk at 1000, 1115 and 1700, on Mondays to Saturdays. On Sundays a more frequent service is provided from Qawra with departures at 0915, 0930, 1000, hourly from 1230 until 1530 and 1600. Return timings from Marsaxlokk are at 1000, 1045, 1200, 1250, 1310, then hourly between 1400 and 1700. The early journeys from Marsaxlokk return direct to Mosta (Targa Gap), instead of serving Tarxien, Attard and Lija.

Route 627 Qawra – Sliema – Three Cities – Marsaxlokk

This is the other "Direct" route between the same towns. However, on leaving the Bugibba and Qawra area, this service covers the main coast-road around the north of the island to Sliema via all the townships served by route 68. After passing along Sliema Front the bus turns right, away from the coast crossing the main Triq d'Argens served by all the routes to the Sliema area outward from Valletta, and climbs up to join the Regional Road to Marsa from where the bus turns off to go through Paola and Vittoriosa, past De La Salle College, to the Hompesch Arch and Zejtun and on to Marsaxlokk. The return journey omits the Vittoriosa leg, instead going through Tarxien to reach Paola.

On weekdays and Saturdays this route operates from Qawra at 0930, then hourly from 1100 until 1500, with a return journey from Marsaxlokk at 1045, then hourly from 1215 until 1615, via Tarxien and Paola to reach Sliema, and then the coast-road to Qawra. The Sunday timetable provides services at 0935, 1015 and then hourly between 1100 and 1500, with journeys back from Marsaxlokk at 1120, and hourly between 1230 and 1630.

ROUTES to ZURRIEQ, QRENDI, HAGAR QIM and WIED iz-ZURRIEQ

Route 32 Valletta – Marsa – Luqa — Kirkop – Safi – Zurrieq

After covering the main route to Marsa the bus makes almost a 360 degree circuit of the one-way system and so heads out to Luqa, passing Malta's golf-course which adjoins Marsa Sports Club with its tennis, cricket and rugby football pitches. A left turn into Triq Valletta leads to a long but gentle hill through the light industry units where the Scarnif bus bodying factory is situated, and then Luqa village is reached. Originally an agricultural settlement established in 1634, Luqa nowadays gives its name to Malta's International Airport. The area suffered badly during the Second World War, being reduced to rubble by enemy aircraft. The town is now rebuilt and is proudly affluent.

The bus passes the headquarters of Air Malta, skirts the western edge of the Airport, passing through a tunnel beneath the runway, before negotiating some narrow streets through the attractive villages of Kirkop and Safi. Safi is the smallest of the villages in the area, with tall palm-trees and a distinctly Moorish influence to the design of the houses, especially their windows and doors. The Xarolla Windmill was used until about twenty years ago, but during a strong gregale the sails and mill were damaged beyond repair.

Zurrieq is the largest village in the south of Malta. Its name is derived from the Arabic for "blue", and so is associated with the clear waters of the Blue Grotto. Its church, dedicated to St Catherine of Alexandria, dates from 1632 and boasts six magnificent paintings by Mattia Preti, who designed the vault of St John's Co-Cathedral in Valletta. Around the town are three Roman towers and a number of smaller historic chapels, those at Hal Millieri being the most famous. These two chapels have been restored during the past forty years and are prime examples of the simplicity of mediaeval Christian worship.

This route operates from 0600 until 2200 from Valletta, and with route 34, there is a fifteen or twenty minute frequency, with most journeys being 34, including all evening departures.

Route 33 Valletta — Marsa Industrial Estate – AFM Barracks or Karwija — Zurrieq

This route operates only outward from Valletta at 0620 and 0635, primarily as a workers' service on weekdays only. After the Airport tunnel, the route serves roads parallel to the southern edge of the airport to the Barracks which are not otherwise served by routes buses.

Route 34 Valletta – Marsa – Luqa – Kirkop – Safi – Nigret — Zurrieq

This route basically follows the same path as route 32, but most outward journeys travel via the residential area of Nigret before arriving at the Zurrieq terminus. The route operates every 30 or 40 minutes from Valletta between 0700 and 2200 and from Zurrieq from 0530 until 2100. Late journeys to Valletta run via the Airport terminal, leaving Zurrieq on weekdays at 2030 and 2100; on Saturdays at 2100 and on Sundays at 2040, 2100 and 2130.

Route 35 Valletta – Marsa – St Vincent de Paule Residential Home — Mqabba – Qrendi

After following the usual route to Marsa, this route serves the Marsa Industrial Estate along Triq L-Ghammieri, then joining Triq Hal-Qormi at the junction overlooked by the St Vincent de Paule Residential Home. The bus soon passes the premises of Zinnu, the well-known bus repairer and coach-builder, as it approaches the roundabout close to Malta International Airport at Luqa. The bus turns right under the runway tunnel and right again at the next roundabout for Mqabba. It soon passes the large limestone quarries of Tad-Dawl. This limestone is used for building Malta's distinctive stone houses. Mqabba is soon reached and after negotiating two roundabouts the bus approaches Qrendi and

FBY 683 is a Bedford YMT/Plaxton Supreme III coach and obligingly waits for a photo call by the Xarolla windmill in September 2009, while working route 34 to Zurrieq.

turns right, up to the town square, which, as in so many other villages, is overlooked by a church, the Church of St Mary. The nearby Cavalier Tower, the only octagonal tower on the island, was built in the sixteenth century. In times of danger the local folk would take refuge in the tower and pour boiling pitch and throw stones from its flat roof onto the attacking forces. This service skirts the village centre of Luqa on the outward journey, but serves the village centre on the return journey to Valletta. It operates from Valletta from 0600 until 2130 and from Qrendi from 0530 until 2030, every twenty or thirty minutes.

Route 38 Valletta – as route 32 to Marsa — Luqa – Zurrieq – Wied iz-Zurrieq – Hagar Qim – Qrendi – as route 35 to Marsa — Valletta

Route 138 Valletta – as route 35 to Luqa — Qrendi – Hagar Qim – Wied iz-Zurrieq – Zurrieq – Luqa — as route 32 — Valletta

Route 38 follows the same roads as route 32 as far as the first roundabout after the Airport tunnel, then skirts Kirkop along Triq Valletta and turns left to serve the main bus terminus in Zurrieq. Travelling clockwise from here, the bus travels down the undulating and picturesque coast-road to Wied iz-Zurrieq. Passengers alight here for the walk down to the pretty fjord in the precipitous cliffs from where a boat ride can be taken to the Blue Grotto and the caves formed under the cliffs. The bus now heads up the gentle ascent along the coast road to the bus-stop for Hagar Qim. Hagar Qim, and the nearby Mnajdra, are the sites of the most important Neolithic temples on Malta, built about 3600 BC, i.e. one thousand years before Stonehenge. Hagar Qim stands on the cliffs which govern this stretch of coastline. It is a temple with several entrances and an intriguing external shrine. The largest stone weighs 20320kg and is over six metres in length. Many of the traditional

Malta and Gozo Buses

DBY 401 is another venerable route bus still in operation. Its chassis dates back to 1932 and its Debono body was built in 1952. The vehicle is seen on the rough, rural road between Qrendi and the bus stop for Hagar Qim Temples on route 38.

'fat figures' of Maltese history (carved stone statues of grossly overweight deities), now in Valletta's museums, were discovered here. There are three temples at Mnajdra, tucked away in a hollow, and, as at Hagar Qim, they offer a wonderful view of the bird sanctuary isle of Filfla out to sea.

The bus then heads back inland along narrow rural lanes past fields which are abundant with poppies during the spring, distinctively high dry-stone walls and farms, soon arriving at the bus terminus at Qrendi. Route 138 operates over the same roads but in an anti-clockwise direction between Qrendi and Zurrieq.

Route 38 operates hourly from Valletta from 0915 until 1615. Route 138 also operates hourly from Valletta between 0945 and 1545. The two routes thus provide a thirty minute frequency on the roads to Hagar Qim and Wied iz-Zurrieq.

Route 39 and Route 36 Valletta – Marsa —Ingriet – Luqa

This departure at 0610 operates on weekdays only and returns to Valletta at 0700, as route 36.

Route 132 Zurrieq – Valletta

This "Express" route was introduced on 2nd May 2007 to provide quicker access to and from the capital at peak times. It operates every half-hour — between 0650 and 0920 from Zurrieq, and between 1640 and 1610 from Valletta — with just six nominated stops en route. Low-floor vehicles are always assigned to the service.

ROUTE 40 to ATTARD

Route 40 Valletta – Msida – Birkirkara – Balzan – Lija – San Anton Gardens – Attard (circular)

Balzan, Lija and Attard are known as the "Three Villages" and boast of being "the places to live" on the island. They are quiet villages, with no industry, just a few shops but attractive houses and many *palazzi*, the patrician homes of the oldest Maltese families.

The bus takes the main route via Msida to Birkirkara, and at the junction where routes 43 et seq. turn right, route 40 continues straight ahead along Triq il-Wied. Birkirkara merges imperceptibly into Balzan hereabouts. The bus then turns right through the village, soon to return to the main road at the roundabout on the edge of Lija. Within the confines of this roundabout the route has its own bus-stop. The bus passes through Lija which is considered to be the most chic of the Three Villages and certainly has the greatest number of *palazzi*. Its Baroque parish church, the Church of the Saviour, was built in the late 1690s and has one of the earliest domes built on a Maltese church. Lija is noted for its *Festa* and fireworks display on 6th August which is the largest on the island.

After Lija, the bus negotiates the narrow streets of Balzan down to the main square which is dominated by the Parish Church of the Assumption, erected in 1665 when the village became an independent parish. The bus then passes by San Anton Palace and Gardens. The President of Malta is the most celebrated resident of the Three Villages, for, since 1974, his official residence has been San Anton Palace. The palace was originally built, in 1620, merely as a private country residence for the French knight Antoine de Paule who used it as his summer home. Once appointed as Grand Master he extended this residence into a palace, because he considered the journey from Valletta to Verdala Castle near Dingli (q.v. route 81) too irksome. Work began on these extensions in 1623 and the palace became the home of later Grand Masters, too.

The palace is not open to the public, although its terraces are. The gardens themselves are one of Malta's highlights. They are planted with trees, such as palms, pines and jacarandas and the paths are laid out in a formal grid pattern, with fountains and a pond at the intersections. The gardens have been open to the public since 1882.

Attard is soon reached and the bus goes along Triq il-Linja, a road running alongside the path of the former Malta railway for most of its length. Turning sharp right into Triq V. Vassallo, the bus passes some attractive houses and gardens. The Renaissance parish church of St Mary was designed by Tommaso Dingli in 1613; it was planned in the form of a Latin cross, has three domes painted in a deep pomegranate red and a pair of belfries. Its campanile was a 1718 addition.

The bus by now is heading back towards Valletta but serves Triq il-Mosta and a short section of "against the flow" travel along Triq A. Schembri, before going along the very narrow Triq San Antnin where passengers can alight for easy access to the Gardens. Close by is Villa Bologna behind the high walls on the opposite side of the road which was built by a Maltese aristocrat, Nicola Perdicomati Bologna in 1745 as a wedding gift for his daughter. It is one of the island's biggest villas, and is still in private ownership. The bus now travels down the narrow roads in Balzan, as it did on the outward journey, but once past the main square, it travels along other residential roads, omitting Lija, and rejoins the outward route along Triq in-Naxxar.

The route operates from Valletta from 0600 until 2200 and from Attard from 0530 until 2100, with departures every twenty or thirty minutes.

ROUTES serving BIRKIRKARA and SAN GWANN

Route 41 Valletta – Msida –Tal Qroqq – Kappara – Ta'-Zwejt – San Gwann (Naxxar Road) – Imrabat – Valletta (circular)

After following the usual route to Msida roundabout, the bus climbs up to the Regional Road. Passing Malta University on the left and the national swimming pool on the right, the bus turns left at the next roundabout and travels along Kappara Hill through San Gwann, eventually turning left onto the main road to Naxxar to the outermost point of the route in Triq V. Borg Brared. The return journey rejoins the main Naxxar road, descending to, and passing under the Regional Road, continuing straight ahead along Triq Birkirkara. The bus passes through the Savoy and Gzira area and turns right off Triq d'Argens, climbing up Triq tas-Sliema, turning left into the Regional Road, and then via Msida, back to Valletta.

Departures from Valletta are from 0600 until 2200 and from Ta' Zwejt from 0530 until 2130, every 20 or 30 minutes.

Route 141

This is an evening variation via Msida, the Regional Road, Triq Sliema, Triq d'Argens, then the main Naxxar road through San Gwann to the Ta' Zwejt terminus, returning the same way, i.e. the daytime inbound route of route 41.

Route 42 Valletta – Msida (Triq d'Argens) – Imrabat—Ta'Giorni – San Gwann Industrial Estate – Birkirkara Parish Church

The bus once again takes the usual route to Msida, then along Rue d'Argens and Triq Birkirkara to the Regional Road where it makes a virtual 360 degree circuit of the gyratory to serve the Ta' Giorni area. The bus turns right onto the main Naxxar road and just before the Ta' Zwejt terminus of route 41, it turns left to serve the San Gwann Industrial Estate, calling at the bus stop at the top of the approach road to the Mater Dei Hospital, then turning right onto Birkirkara Bypass, and then left along Triq Tumas Fenech (which is not served on market-days) to the terminus in front of Birkirkara Parish Church.

As the parish became wealthier during the eighteenth century through increasing support from benefactors, this church, dedicated to Saint Helena, was constructed and celebrated its first *Festa* in 1738. It is considered to be the finest example of Maltese Baroque architecture with an ornate interior of frescoes, carvings and paintings.

Departures from Valletta from 0600 until 2200, and from Birkirkara Church from 0530 until 2115, every half-hour (routes 141 and 142 included).

Route 142

This is an early morning and evening variation which follows route 141 exactly, but extends to Birkirkara Church, returning the same way.

Route 71 Valletta – Floriana – Hamrun – Santa Venera – Fleur de Lys – Birkirkara Station

This route reaches Birkirkara by an altogether different route from 42 and in so doing serves the southern area of the town. The terminus is close to the site of the town's former railway station on the long-defunct Malta Railway.

From Valletta, the bus serves the main road through Hamrun and Santa Venera where the remains of the Wignacourt Aqueduct run parallel to the road for about eight hundred metres. This aqueduct was built in the early seventeenth century as part of the waterway which brought channelled spring water from Girgenti on the coast into Valletta. At the Fleur de Lys roundabout, the bus turns right along Triq Fleur de Lys for most of its length

The terminus at the Old Station at Birkirkara is seen with three route buses on lay-over. From left to right, Bedford YRQ, FBY 737, King Long EBY 572 and Maltese f/c (Baileys)/Schembri DBY 365, all operating route 71 from Valletta.

to the terminus near the Old Station. This terminus is only a few hundred metres from the junction with the main road through Birkirkara covered by the routes to Mosta and the north-west of the island. Departures from Valletta are from 0620 until 2200 and from Birkirkara from 0530 until 2130 every ten or fifteen minutes.

Route 73 Valletta – Hamrun –Psaila Street – Birkirkara – Lija Church

This route reaches Birkirkara along the same roads as route 157 (q.v.). At the roundabout at the junction with the Birkirkara bypass, the bus turns left on the outskirts of Lija. From here it is only a short distance along Vjal it-Trasfigurazzjoni to the terminus at Lija church.

There are three daily morning departures from Valletta at 0630, 0730 and 0830, each returning from Lija thirty minutes later.

Route 74 Valletta – Hamrun – Psaila Street – Birkirkara – Balzan

This route, too, follows route 157 to Birkirkara, but the bus then continues along Triq il-Wied (as does route 40) towards Balzan, soon turning left along Triq il-Ferrovijal l'Qadima to the terminus at the Corinthia Palace Hotel.

There are just two daily departures from Valletta at 0700 and 0800, returning from Balzan half an hour later.

ROUTES 43, 44 and 45 to MELLIEHA and CIRKEWWA

The many routes which serve the north-western and northern resorts are obviously some of the longest services on the island, with route 45 to Cirkewwa being the longest of all (over twenty-four kilometres) and being timetabled to take seventy minutes from Valletta. While the frequencies on these routes may not be quite as great as those to the Sliema area, a ten-minute interval is the norm for much of the day. From Valletta buses pass through Msida then make the steady climb up to Birkirkara. At the western edge of the town buses turn right, skirting Balzan and Lija, passing Mosta Technopark. Mosta is soon reached, via the lengthy one-way system with all the outward routes going along Triq l'Indipendenza and Triq il-Kbira to reach Mosta Dome, the usual name by which the parish church of the Assumption is known. Begun in 1833, when Mosta was still a small village, the Dome took twenty-eight years to be built. The dome itself was constructed without the use of scaffolding, because it was built over and around the original village church which had become too small for the growing parish and could not be demolished until the new church was finished. On completion the Dome was a source of controversy. The archbishop refused to bless the new church himself, as he did not approve of the circular plan of the church, preferring the traditional Latin form of the crucifix. At that time the Dome was claimed to be the third largest unsupported dome in Europe. Since the 1950s, however, the church at Xewkija on Gozo has taken over third place (see routes 42/43 on Gozo).

The church has a bright interior with sixteen windows, a lantern light and a floor with a geometrical pattern laid in two different marbles. During World War Two, on 9th April 1942, a Luftwaffe bomb crashed through the dome but did not explode. The 300 parishioners at Mass escaped safely. A replica of the defused bomb is now on display in the church.

After the bus-stop at the Dome, the long straight Triq il-Kunstizzioni climbs up towards the roundabout at the head of the Targa Gap. Straight on at the roundabout, the bus soon begins the descent, crossing the Victoria Lines and swinging right round the sharp hairpin bend halfway down. On the distant coastline the hotel skyline of St Paul's Bay, Bugibba and Qawra can be seen. The bus now passes through low-lying arable fields to Burmarrad. Nearby, on the hillside of Gebel Ghawzara, stands the church of San Pawl Milqi (St Paul Welcomed) where, tradition has it, Paul, Luke and other survivors of the shipwreck in AD60 were cared for by Malta's Roman governor, Publius, whom St Paul converted to Christianity. The present church was built in 1620. Archaeological discoveries have shown that farming was carried out as early as the second century BC in the immediate surrounding area. The next roundabout is at the junction of the bypass around St Paul's Bay to the left, and, on the right, the main road coming in from the northern coast-road from Sliema.

An individual survey of each route now follows, with a route description continuing from this roundabout at the end of the bypass.

Route 43 Valletta – Msida – Birkirkara – Mosta – St Paul's Bay – Mistra – Mellieha

At the roundabout the bus continues straight ahead, along the main road through St Paul's Bay, which until the turn of the twentieth century was a peaceful fishing village. But as the Valletta businessmen became more affluent, they decided to purchase their "summer houses" in the area —"away from Sliema", they would say – and so the popularity of the village grew. Wignacourt Tower overlooks the bay. Built in 1609 it was the prototype for other similar towers on the island.

This view of EBY 529, taken in August 2008, shows the countryside at Ghadira Bay to its best. The Red Fort is clearly visible on the summit of the hill in the background. Route 48 links the holiday resorts of Bugibba and Qawra with the Gozo ferry terminal at Cirkewwa.

The narrow road through the centre of the village soon opens out onto the shoreline of Pwales Beach. Here the bus turns right, skirts the sea edge through Xemxija, "the sunny place", then climbs a steep hill to Mistra village, a growing holiday centre, and then descends Mistra Hill before tackling the more fiercesome Selmun Hill with its double hairpin up to Mellieha Ridge where at the roundabout, the Mellieha Bypass starts.

Continuing straight ahead here, the bus passes through the outskirts of Mellieha, then fights its way through the village centre, down a steep hill always lined with parked vehicles, to the terminus near the police station. Mellieha stands on the ridge overlooking the northern coastline and Gozo beyond. The village was one of the ten original parishes of 1436 on the island. Despite its lofty position, it was a favourite target for corsairs, seventeenth century privateers in the southern Mediterranean, who could easily land at Ghadira Bay, then seize the local inhabitants and sell them as slaves and by the mid-1500s the village had been depopulated. It was not until the security provided by British naval vessels patrolling offshore at the start of the nineteenth century that the town began to flourish. The parish church of the Nativity of Our Lady dominates the skyline when viewed from Ghadira Bay.

Route 44 Valletta – Msida – Birkirkara – Mosta – St Paul's Bay — Mistra – Mellieha (Ghadira Bay)

This route continues past the terminus of route 43 in Mellieha and then negotiates another double hairpin bend on its descent, travelling between dry-stone walls and oleander bushes, towards the coast. Ghadira Bay, one of the best sandy beaches on the island is soon reached. The terminus is at the nature reserve situated halfway along the dual carriageway fronting the sea. This nature reserve is formed on the old saltpans of the area – "Mellieha" means "a salt dealer".

Mid-morning in September 2009 and EBY 565 attacks the first of the hair-pin bends, as it descends from Mellieha to the coastline at Ghadira Bay. The parish church dominates the scene on the hill-top and oleander bushes are blooming in abundance in the foreground.

Ghadira Bay is the scene of many Roman wrecks; it was the landing-place of Don Garcia de Toledo, the Spanish viceroy of Sicily when he brought his relief force to the island during the last days of the Great Siege, on September 7th, 1565: Napoleon chose the bay as one of his landing points, when he overran Malta in 1798 and drove out the Order of St John. The British forces built pillboxes and trenches around the bay during World War Two to protect the area against enemy landing-craft.

Departures from Valletta on routes 43 and 44 are infrequent, providing just a handful of journeys, particularly in the evenings, to supplement route. An interesting variation of route 44 is the service from the forecourt of the Mellieha Bay Hotel at 1015 and 1630, interworked with the Popeye Village service, route 441.

Route 45 Valletta – Msida – Birkirkara – Mosta – St Paul's Bay – Mistra – Mellieha – Cirkewwa

Once at Ghadira Bay, route 45 tackles the final hill on the road to Cirkewwa, climbing up Marfa Ridge past the Mellieha Bay Hotel. The Red Tower stands on this ridge. It is a terracotta-coloured building built by the Grand Master Lascaris in 1649 to protect the northern part of the island and also to act as one of the warning towers across the island. From the ridge-top there is an immediate descent along Triq il-Marfa. Eucalyptus and mimosa trees line the first part of the descent before a bleak section of road brings the bus to the coast again. After a left turn, with views across to Comino and Gozo, the bus passes the Paradise Bay Hotel and arrives at the terminus at Cirkewwa, the main ferry terminal for Gozo.

Departures from Valletta are from 0535, every ten or fifteen minutes, reducing to fifteen or twenty minutes in the evening. The last bus from Valletta is at 2040 and from Cirkewwa at 2010. This is the longest route on the island taking seventy minutes each way.

Route 145 Valletta – Msida – Birkirkara Bypass – Mosta – St Paul's Bay – Mellieha — Cirkewwa

This is a weekday only variation of the main route 45 which, on reaching Msida, goes via Birkirkara Bypass. It operates every hour from Valletta between 0810 and 1710. It is thought that the minor deviation via Birkirkara Mill has been withdrawn, though route indicators are still in situ to direct drivers via Triq Ganu in Birkirkara.

Route 50 Ghadira Bay – Armier Bay

This is an hourly summer-only service which provides easy access to beaches close to the Marfa Ridge. It operates in conjunction with route 441, and leaves Ghadira Bay hourly between 1015 and 1615 after returning from Popeye Village. The bus climbs up to the top of Marfa Ridge, turns right so that the Red Tower is directly behind, along a very rough road, passing the lane leading to the Ramla Bay holiday complex, then turning left, making a long straight descent to Armier Bay with its café and sandy beach. Further on, but unserved by route 50 is Ahrax Point, the northerly tip of Malta.

Route 50 Valletta – Msida – Birkirkara – Mosta – St Paul's Bay — Mellieha – Armier Bay

This Sunday only version of route 50 operates hourly from Valletta between 0900 and 1600, returning from Armier Bay between 1000 and 1900 during the high summer.

Route 441 Ghadira Bay – Popeye Village.

This is a shuttle service operating hourly during the main summer months, but only every two-hours for the remainder of the year, between 1000 and 1600. There is no Sunday service. The distinctive destination board shown in a photograph in the first edition of this *Bus Handbook* has been discontinued, since the introduction of the 441 route number. From Ghadira Bay the bus merely travels along the coast-road in the direction of Mellieha to the first roundabout, where it turns right on the Mellieha Bypass for a few hundred Metres, before bearing right along a narrow lane to Sweethaven (or Popeye Village), past some horse-riding stables and over some "sleeping policemen". Popeye Village was created as the set for the 1980 film *Popeye* which starred Robin Williams.

DBY 308 is an AEC Mercury (Zammit) Gauci body built in 1966. It is seen at the ferry terminus for Gozo working 48 from Qawra.

ROUTES serving BUGIBBA and QAWRA

Route 49 Valletta – Msida – Birkirkara – Mosta – St Paul's Bay – Bugibba – Qawra

Having reached the roundabout on the outskirts of St Paul's Bay, the bus continues straight on, soon bearing right to the waterfront near to Gillieru Pier. The bus turns right passing St Maximilian Kolbe Church, and travels along the esplanade, lined with palm trees, bearing right by the pier at Bognor Beach into Bugibba's Islet Promenade. This road is lined with shops, snack bars, souvenir stalls, a dodgem track, and sports facilities. Bay Square is alive with cafés and gift shops. Turning right after passing the Oracle Casino and Dolmen Hotel, the bus soon reaches the terminus at Qawra. In 2001, a bus-lane along Islet Promenade was created for buses travelling from Qawra in the direction of St Paul's Bay, along the section of the promenade which is otherwise one-way for all traffic. During the summer of 2009, much of the promenade was closed to all traffic between 2000 and 2400 on Friday, Saturday and Sunday evenings. Buses in service to the resorts had to make the long anti-clockwise journey from the roundabout at the end of the St Paul's Bay by-pass around the Qawra peninsula.

Bugibba is a popular holiday centre and boasts a large number of time-share apartments and 2-, 3- and 4-starred hotels. Qawra, too, is a popular resort with fine hotels, many overlooking Salina Bay to the east. Route 49 operates from 0645 until 2200 from Valletta, and from 0530 until 2130 from Qawra, with departures at 7, 10 or 15 minute intervals. On summer Sundays the service is more frequent.

Routes 58 and 59 Valletta – Malta University – Mosta – Bugibba – Qawra

Route 58 takes an altogether different way from the others in this group, once it arrives at Msida. It climbs up to the Regional Road and continues along the Birkirkara by-pass past the extensive grounds of Malta's University and the Mater Dei Hospital, with the narrow streets of Birkirkara on the left and the San Gwann Industrial Estate on the right. At the roundabout at the junction with Triq in-Naxxar, the bus turns right, thus rejoining the main road and follows route 49 to Qawra. Inbound buses serve the Hospital and University sites.

Route 59 serves Naxxar en route to Mosta and after the roundabout at St Paul's bay, Qawra seafront, this omitting Bugibba.

FBY 775 has a Dodge chassis with Zammit bodywork and is seen travelling along the esplanade in Bugibba on route 49 from Valletta to Qawra.

On Saturday afternoons and all day on Sundays the 59 is re-numbered 159 (q.v.) and runs via Psaila Street, as route 157 does not operate at these times.

Route 153 Mosta – Birkirkara – Msida — Valletta

This is the other "Express" route, which, as route 132, was introduced on May 2nd 2007. Peak time departures are every thirty minutes between 0650 and 0920, with, again, six designated stops en route. It is believed that the evening departures are no longer operated.

Route 48 Qawra – Bugibba – St Paul's Bay –Mistra – Mellieha – Marfa – Cirkewwa

This is one of the "Direct" routes on the island, providing an easy link with the Mellieha area and the Cirkewwa ferry terminal for the holidaymakers and residents of Qawra and Bugibba.

The route simply leaves Qawra along Islet Promenade, bearing left into Triq il-Halel, thus avoiding the bus-lane, and serves some of the hotels and apartment blocks away from the seafront in Bugibba. It soon arrives at the roundabout by the St Paul's Bay bypass, turns right onto the main route all the way to Cirkewwa.

Route 48 operates from 0830 until 1700 from Qawra, and from 0910 until 1750 from Cirkewwa between October and March, with summer timings departing from Qawra until 1830 and from Cirkewwa until 1915. Departures are every 20 or 30 minutes.

Route 449 Bugibba – Coast Road – Regional Road – Valletta

This is an inbound "Express" route which operates twice daily at 0935 and 1035.

Route 645 Sliema Ferry – St Julians – Paceville – Splash and Fun Park – St Paul's Bay– Mellieha – Cirkewwa

This is another "Direct "route which covers the main northern coast road from Sliema (q.v. routes 60 to 68) past the outermost terminus of the Sliema routes (route 68) at Bahar ic-Caghaq and onwards to Salina on the outskirts of St Paul's Bay. Passing through Kennedy Grove, a popular grassy picnic area, the bus soon reaches the roundabout at the junction of the St Paul's Bay bypass. The bus turns right and continues to Cirkewwa, as if on route 45. Departures from Sliema are from 0700 until 1830 and from Cirkewwa between 0900 and 2205 during the summer. During the winter, timings are 0800 until 1815, and 0900 until 1910, respectively. The frequency is generally every thirty minutes. In winter, morning journeys towards Sliema divert via Qawra bus terminus.

One of the principal routes from Qawra to the capital is route 58 which now serves the Mater Dei Hospital and University grounds on its inward journeys. FBY 803, a Bedford SB with Barbara bodywork is soon to take up position on the stand for its next journey.

ROUTES to GOLDEN BAY

Route 47 Valletta – Msida – Birkirkara – Mosta – Ta'Hagret Temples – Mgarr —Ghajn Tuffieha – Golden Bay

This route serves the rural and peaceful area north-west of Mosta towards the coast. From Valletta to Mosta, the route follows exactly the same roads as the routes to Mellieha (q.v. routes 43, 44, 45). On leaving the bus stop at Mosta Dome, all outbound buses follow this comparatively new variation of the route: immediately after Paramount Garage the bus turns left along Triq it-Tama to serve Ta 'Zokkrija estate, then left down to a roundabout where it makes an almost 360 degree circuit before heading westwards towards Mgarr. Passing the southern extremities of the Victoria Lines at Falka Gap, the bus passes though Malta's least populated region, an agricultural area famed for its strawberries. On the approach to the small town of Mgarr, the bus passes the Skorba Temples, which are prehistoric sites dating from about 3600BC. These Temples and the Ggantija remains on Gozo (q.v. route 64 on Gozo) are credited as being the oldest free-standing structures in the world.

Mgarr is a quiet farming community which was first settled in the mid-nineteenth century. Its tall parish church has an incongruously small dome, which has been likened to an egg in an eggcup. The church was built by donations from the local parishioners, as so often on the islands. However, in this case, the poor farming community could afford to bring only produce from their farm holdings, which the parish priest then sold to raise cash.

Ta-Hagret Temples are close to the village square and also close by, is Mgarr Air Raid Dungeon which is reached through the Il-Barri Restaurant and which has been restored to provide a faithful evocation of how the Mgarr residents lived under shelter during the ferocious attacks on the island during 1941 and 1942.

In the Square the bus turns sharp right and proceeds uphill through the village along Triq Sir Harry Luke, named after a former Governor who wrote a detailed history of the islands. It turns left at the next T-junction, before descending through pleasant countryside past the Roman Baths which still show how sophisticated the Romans were, with changing rooms, lavatories, an early form of sauna, and heated pools. The main road is lined with oleander bushes on the descent to the traffic island at the junction with the road from St Paul's Bay. Turning left, the bus approaches Ghajn Tuffieha Bay but then sweeps right, down the straight hill to Golden Bay. These two bays are separated by a gentle peninsula and are acknowledged as two of the island's best beaches. Nearby are a woodland tourist village and horse-riding centre.

The village of Manikata on a nearby hillside overlooking the Pwalles Valley is served nowadays by an anti-clockwise circuit begun by turning right at the traffic island mentioned in the previous paragraph, then second left up to Manikata and then down to the Golden Bay terminus.

Return journeys from Golden Bay serve Ta 'Zokkrija only hourly. Otherwise the route covered into Mosta is via Triq il-Kbira from the roundabout mentioned in the first paragraph.

Route 47 operates from 0900 until 1937 from Valletta; from Golden Bay departures begin at 1000 and end at 1830, (basically) every thirty minutes. Earlier journeys are provided by route 52.

Route 46 Mgarr – Mosta – Birkirkara — Valletta

This service, a short-working from Mgarr, via Ta 'Zokkrija estate, operates in the morning rush-hour on weekdays.

Route 52 Valletta – Msida – Birkirkara – Mosta – St Paul's Bay – Manikata – Ghajn Tuffieha – Mgarr – Valletta (circular)

This early morning route has departures from Valletta at 0600 (Monday to Saturday) 0700 and 0815 (both daily). The route is merely a combination of the main route from Valletta to St Paul's Bay, from where the bus follows route 652 along the Pwalles Valley and then turns right to travel up to Manikata, before arriving at the Golden Bay terminus. Here the vehicle merely resumes its journey back to Valletta by covering route 47 via Mgarr. This is, by far, the longest circular route on the island, lasting almost two hours.

Route 652 and Route 51

Sliema – St Julian's – Paceville – St George's – St Andrew's — Splash and Fun Park – Bahar ic-Caghaq – Qawra terminus – St Paul's Bay – Ghajn Tuffieha (Golden Bay)

Route 652 is another "Direct" services on the island. As so many of the other Sliema-based routes, it follows the coast road through the holiday resorts and out to the Splash and Fun Park. The main coast-road eventually approaches Qawra and near to Kennedy Grove, the bus bears right up to the main road through Qawra overlooking Salina. It soon arrives at the bus terminus at Qawra, from where it passes along Bugibba Islet Promenade before turning left away from the sea, to head to the roundabout at St Paul's Bay. Turning right here, the bus travels through the narrow streets of St Paul's Bay and at the roundabout at the head of Pwales Beach close to Xemxija, the bus heads off down the fertile Pwalles Valley where farming land and greenhouses are in abundance. The bus eventually joins route 47 at the traffic island near to Ghajn Tuffieha where it, too, sweeps right, down to Golden Bay.

Route 652 operates from 0850 to 1800 from Sliema and from 0900 until 1905 from Golden Bay, every 20 or 30 minutes (depending on the time of day) during the summer. In winter, the timings are only every half-hour – between 0845 and 1615 from Sliema, and from 0935 until 1705 from Golden Bay.

Route 51 is a short-working which covers the Bugibba/Qawra to Golden Bay section in the early evening, to relieve overcrowding on the main 652 service, leaving Qawra at 1830, 1900 and 1930, though timings are dependent on seasonal demands and loadings.

EBY 602 is a BMC Falcon seen working a timing of route 47 to Golden Bay which serves the hillside village of Manikata.

ROUTES to NAXXAR, GHARGHUR and SANTA MARGARITA

Route 54 Valletta – Msida — Birkirkara — Naxxar Schools (schooldays only)

There are six morning departures on this route at 0715, 0735, 0745, 0752, 0810 and 0830. Buses then run light to Qawra Bus Terminus to take up services from there, primarily route 49. Afternoon journeys include departures from the schools at 1230, 1320, 1400 and 1500.

Route 55 Valletta – Msida – Birkirkara – Naxxar – Gharghur

For much of its way from Valletta this route follows the same roads as routes 43 et seq., all the way to Birkirkara and past Lija cemetery. However, at the roundabout on the approach to Mosta Technopark, the bus turns right and begins the steady ascent of Triq Tal-Labour which leads directly to Naxxar town centre where it circles the parish church and travels along Triq il-Parrocca to the roundabout. Here it turns left and almost immediately right, along Triq Gharghur. On the approach to Gharghur village, the bus bears right and then left into Vjal ir-Repubblika, continuing into Triq l'Oratorju where the terminus is in front of the church.

The return journey serves Triq Bartolimew and Triq San Gwann, then as the outward route. Route 55 operates from 0645 until 2200 from Valletta, and from 0615 until 2045 from Gharghur every 15, 20 or 30 minutes.

Route 56 Valletta – Msida – Birkirkara – Mosta – Naxxar – Xwieqi —Gharghur

This route follows exactly the same roads to Mosta Dome as routes 43 et seq. Instead of turning left by the Dome, the bus continues straight ahead, past the Dome's main façade and heads out along the one-way Triq il-Kbira which leads directly into Vjal il-21 ta'Settembru. At the junction just past Naxxar Community Centre, the bus turns left into Naxxar town centre and approach to Mosta, the bus travels along Triq Sant Antnin, Triq Barrieri and Triq Oratorju throughout the one-way system.

Route 56 operates – mainly — every hour on the hour from Valletta from 0600 until 1700 with later journeys at weekends.. Naxxar stands to the northeast of Mosta on a hill which overlooks the surrounding countryside and the Victoria Lines. Naxxar church, dedicated to the Nativity of Our Lady, is a Baroque edifice. The Trade Fairground lies to the west of the town square where international and home-based fairs are regularly arranged.

Route 157 serves the pleasant residential area of Santa Margarita to the north of Mosta, and unusually reaches Mosta via Hamrun instead of the more usual route via Msida. FBY 701 a GMC f/c with Brincat bodywork, pauses in Triq Dura.

Nearby, the Palazzo Parisio is a finely decorated C19th "stately home", now much favoured as the venue for weddings and parties.

The Victoria Lines are a defensive barrier of fortifications about sixteen kilometres in length from Fomm ir-Rih Bay on the coast south of Mgarr to Fort Madliena on the northern coast close to the White Rocks Holiday Complex near Pembroke. These Lines were planned as an outer defensive barrier for Valletta. During the mid-nineteenth century the British army became increasingly worried about the development and increasing power of artillery fire. Thus attacks on the island would not necessarily involve an immediate assault on Valletta itself, rather an attack from further away. Army strategists proposed the idea of developing the natural feature of the Great Fault which cuts off the north of Malta from the rest of the island. Thus the Victoria Lines, merely walls about two metres high, were built to provide cover for the army firing on attackers below. Work on building these defences was in progress from 1874 until 1897, being completed in Queen Victoria's Diamond Jubilee year – hence their name. A subsidiary network of defences was also built on the ridge at Dwejra near Mosta.

Route 157 Valletta – Hamrun – Santa Venera – Birkirkara – Mosta – Santa Margarita (circular)

Buses travel from Valletta via Hamrun, turning right near the main Post Office, and then travelling along Triq Kappilan Mifsud, over the Regional Road, along Triq il-Ferrovija and past the terminus of route 71 in Birkirkara. The bus then joins the main road to Mosta, where it heads out towards Naxxar, turning left into Vjal il-Qalbiena Mostin, before a right turn takes the bus through the residential area of Santa Margarita, soon reaching Triq id-Difiza Civili which overlooks the Victoria Lines. At the roundabout near the Targa Gap, the bus turns left, rejoining the main road into Mosta, and so starts its inbound journey to Valletta. This is an hourly route from 0710 until 1910 on weekdays, and then only on Saturday mornings at the weekend between 0910 and 1210. Return timings from Santa Margarita are from 0945 until 1745 on weekdays and between 0940 and 1240 on Saturdays.

Route 159 Valletta – Psaila Street – Birkirkara – Naxxar – Mosta – Qawra seafront – Qawra

This is a variation of route 59 via Psaila Street operating only on Saturday afternoons and Sundays when route 157 does not operate.

Gharghur is a township close to Naxxar and DBY 338, an Albion Chieftain with distinctive Brincat B45F bodywork, pauses alongside some distinctive "street furniture" in Triq San Gwann in March 2009.

ROUTE 154: NAXXAR LOCAL ROUTES

The Naxxar town route was introduced on 16th January 1995 with a civic ceremony marking the occasion. The first day's operations were provided by Bedford YLQ/Plaxton Y-0583 (now EBY 583), a vehicle owned and driven, appropriately, by a Naxxar resident, Mario Bonavia. The current service is provided Mondays to Fridays (originally Saturdays, too) throughout the year with departures arranged according to the school year. From October 1st to May 31st: 0800 from the bus terminus and 1435 from the Primary School. Between June 1st and July 10th: 0800 from the bus terminus and 1230 from the Primary School. During the summer holidays and until September 30th: 0815 from the bus terminus and 1215 from the Primary School as morning play-groups are arranged at the Primary School during the summer holidays. Evening services throughout the year depart at 1700, 1830 and 1930.

The bus departs from Naxxar bus terminus near the Church of Our Lady of Victory and serves Vjal il-21 ta'Settembru, Triq l-Imraden, Triq il-Qoton, and through to Triq l-Emigranti, from where there is a fine view across to Iklin and Birkirkara. The bus turns right and soon reaches Triq is-Sghajtar, Triq it-Tuffieh and Triq Jean de la Vallette. The bus then crosses Vjal il-21 ta'Settembru and carries along Triq Jean de la Vallette, soon reaching Triq Jules Verne before arriving at the roundabout overlooking Naxxar Gap, where another wonderful view towards the coast at Qawra can be seen. The bus turns right at the roundabout, and travels the length of Triq San Pawl tat-Targa, turning right into Triq il-Parrocca and back to the bus terminus.

Two other special services are provided by the Council, both being seasonal. One summer service on route 154 operates from mid-July to the end of August every Tuesday at 0845 and 1130. From the terminus the bus heads for the northern coast road at Salina, passes the Coastline Hotel and travels along the coast road to the terminus of route 68 at Bahar ic-Caghaq. The bus then returns to Naxxar via Maghtab.. The other seasonal route operates from Naxxar to Ghadira Bay from the first Saturday of July to the last Saturday before the autumn school term starts, leaving Naxxar at 0830 and returning from Ghadira Bay at 1200. Only local residents are allowed to travel on these two routes.

FBY 741 is seen waiting for pupils attending summer school during their holidays. The bus leaves the Junior School in Naxxar soon after midday and then travels around the town on local route 154.

ROUTES 60 to 68 to SLIEMA and beyond

From Msida roundabout all the services travel along Triq d'Argens. But within a few hundred metres , on the approach to Gzira, services divert, either along Triq Princippesa Margarita, or along Triq l-Msida or else keep straight ahead.

Route 60 Valletta – Msida — Sliema Savoy

Route 63 Valletta – Msida – Sliema Savoy – Sliema Ferries – Ta'Xbiex – (Tigne) — Msida – Valletta

Both routes continue along Triq d'Argens, climbing the hill towards Holy Trinity Church, before bearing right into Triq Rodolfu.

Route 60 is the shortest of the routes, terminating at the junction with Triq San Gakbu. Vehicles reverse into the bus-bay at this junction. An interesting feature of the first section of the return journey is that the bus has to travel "the wrong way" along the one-way street.

Route 60 operates from Valletta from 0600 until 2200 and from Savoy from 0530 until 2130. There are nine rush-hour departures from 0600 until 0900; thereafter the services operate hourly until 1700, then half-hourly until 2200.

Route 63 continues past the terminus of route 60, descending Triq Adrian Dingli to the coast-road at St Julian's Point, where the bus turns right and so now heads back to Valletta. The section of the route through Tigne is not covered at the time of writing, as the promontory is currently undergoing extensive redevelopment. Tigne was once the site of a British army barracks and family quarters. The remains of Fort Tigne face Fort St Elmo on Valletta's peninsula, both built to guard the entrance to Marsamxett harbour. Route 63 operates every thirty minutes throughout the day from 0645 until 2020.

Routes 61 to 68 serve Sliema and further areas along the coast; the higher the route number (62 to 68) the further along the coast the outer terminus is. Once past Msida, most of these outbound routes continue along Triq d'Argens and turn right along Triq Msida to reach the sea-front. Inbound buses take a slightly different route from this point back to Msida, by travelling along the edge of Msida Marina. The Strand (known also as Tower Road or as "The Front" by the locals) is a broad dual-carriageway, at the far end of which is the terminus of Sliema Ferries, where passengers can board the Marsamxett Ferry operated by Captain Morgan ferries to travel across the bay to reach Valletta, though there is a steep climb up to the capital from the ferry terminal for Valletta. Here, too passengers

Solaris Valletta FBY 771 is seen in Swieqi, the fashionable locality just south of Pembroke. Route 64 makes a lengthy circuit of the area every thirty minutes.

Malta and Gozo Buses

DBY 408 is an AEC Mercury with Farrugia B40F bodywork. It is seen at the distinctive terminus of route 60 to Savoy where the vehicle has reversed into this narrow street and in this view completely hides the bus shelter at the terminus. As the bus sets off back to Valletta, it travels the "wrong way" along the one-way street.

can join one of the Direct routes from Sliema to popular towns, villages and tourist spots across the island, without the need to travel via Valletta or catch one of the Open-Top services for a tour of the island.

Beyond the Strand, buses pass through a narrow shopping area and soon rejoin the coastline near the Preluna Hotel. St Julian's (San Giljan) is reached where the coastline is indented with small coves, such as Balluta Bay and Spinola Bay. There is a broad esplanade along the sea edge almost as far as Paceville.

Sliema is the main residential area of the island as well as being the major holiday area. The town's early development took place as it was the resort favoured by the residents of Valletta and then became the place where the wealthier citizens of the capital wished to move to. Elegant villas in cream limestone were built and they brought a cachet to the area in the early twentieth century. Gradually the town's outer limits were extended as more and more residents wished to move to the area and Spinola and Balluta were developed. Hotels and apartments have replaced the elegant villas and dominate the waterfront nowadays. At Balluta Bay there is a poolside lido where the Neptunes, one of the island's best water-polo teams plays its matches on Saturday evenings. Paceville is the centre of the island's nightlife.

Route 61 is merely an inbound route from Sliema Ferries to Valletta at 0930, 0950, 1010 and 1100.

Route 62 to Paceville.

At the top of Spinola Hill route 62 reaches its terminus at Paceville, by day an unprepossessing residential area, by night it is the island's most crowded and hectic venue for clubbers. Buses lay over in a side road and pick up in the main road. Opposite the terminus are the bus stops for services to townships further along the coast road, to Qawra and beyond. In addition there are road markings for all the special late-night services. Route 62 operates at 5, 7 or 10 minute intervals throughout the day from Valletta between 0530 and 2300, with similar timings from Paceville terminus.

Night services to and from Paceville

Route 62 Paceville to Valletta operates every half hour between 2330 and 0200 and from Valletta from 2320 until 0220 at weekends, and on weekdays from Paceville between 2300 and 0300 and from Valletta between 2320 and 0220 every fifteen minutes.

On Friday and Saturday evenings buses from across the island provide services to enable the clubbers and holiday-makers enjoy a night out and then travel home in the early hours. These late-night routes have been regularly advertised on the side of some of the route buses. All these following routes depart from Paceville every thirty minutes from 2400 until 0130 in winter, and between 2400 and 0300 in the summer.

Route 49 Paceville – Qawra Bus Terminus – Mosta – Naxxar – Lija Tower – Birkirkara Bypass

Route 118 Paceville – Three Cities – Zabbar – Fgura – Zejtun – Birzebbugia – Ghaxaq — Gudja

Route 134 Paceville – San Gwann – Paola – Tarxien – Santa Lucija – Luqa – Kirkop – Safi – Zurrieq – Qrendi — Mqabba

Route 881 Paceville – Santa Venera – Hamrun – Qormi – Siggiewi – Zebbug – Attard – Rabat — Dingli

Beyond Paceville there are four more routes in the 60s series from Valletta.

Route 64 to Swieqi

Buses continue past Paceville, almost immediately turning right at the traffic-lights onto Signal Road and then sharp left in the narrow roads through Swieqi, an attractive residential suburb. This circular route rejoins the main coast road opposite the St Andrew's terminus and operates every half-hour between 0600 and 2000. The British garrison was stationed in St Andrew's, their barracks being built in the local limestone. It is said that the plans of these barracks were intended for units in India but found their way to Malta instead. Nowadays they are used as private businesses or as government buildings.

Route 66 to St George's Bay

This route was introduced in June 2000 to cater for the extensive hotel developments overlooking St George's Bay in the Pembroke area. After passing through Paceville, the route turns right at traffic-lights onto the Regional Road (hereabouts called Signal Road) and, halfway up the hill to St Andrew's, the bus turns right to make a lengthy circuit passing the Radisson Hotel, Villa Rosa and Hilton Hotel, amongst others, rejoining the main road in St Julian's. This circular route operates every thirty minutes from Valletta between 0710 and 1950, and inbound from St George's Bay between 0730 and 2000.

Route 67 to St Andrew's

This route continues up the hill from route 66's right turn to its terminus on the main road at St Andrew's. This is another very frequent route, operating at 5, 7 or 10 minute intervals throughout the day from 0530 until 2300.

Route 68 to Bahar ic-Caghaq

Buses on this route travel further on past St Andrew's climbing to overlook the coast and Madliena Tower beyond Pembroke and then descending to sea-level past holiday complexes and the Splash and Fun Park to the terminus on the main road at Bahar ic-Caghaq, close to the entrance to the Mediteraneo Marine Park. This route operates every thirty minutes from Valletta from 0615 until 2015.

Other routes in the area:

Route 662 Valletta – Msida – Regional Road — Paceville

This is an express service operating on Friday, Saturday and Sunday evenings only.

Route 667 Valletta – Msida – Regional Road – Paceville – St Andrew's

This express route covers the usual route to Msida, where, at the large roundabout, it climbs the short hill to Regional Road, a fast dual carriageway which takes much of the direct traffic off the urban roads in Valletta, the Three Cities and Sliema. This Regional Road eventually merges into the main road around the north of the island when it becomes Signal Road (see route 66) and the terminus is close by.

The route operates half-hourly from 0635 until 0905 from Valletta, and from Paceville at 0635, 0735, 0805, 0905 and 0935.

Route 671 Valletta – Pembroke — St Andrew's

This route serves all the usual places before reaching Pembroke where, on turning right off the main road just before the St Andrew's terminus, it makes a circuit of the government housing developments close to the firing ranges of the Armed Forces of Malta. There are just two timings at 0630 and 2030 from Valletta, daily.

Route 70 Qawra – Bugibba – St Andrew's – St Julian's – Sliema Ferries

This is another "Direct" route which links the two principal holiday areas on the island along the northern coast-road. Buses from Sliema approach the Qawra terminus via Pioneer Road and Bugibba Square and then serve Qawra seafront. Timings from Sliema are from 0730 until 1910 during the winter or 2030 in summer and from Qawra between 0800 until 1945 in winter and 2230 in summer every fifteen or twenty minutes.

Sliema Local Route

On Monday 28th January 2003 Sliema Local Council introduced a new free bus service for local residents. This circular service (unnumbered) operates from Sliema Ferries at 0800, 0900, 1000 and 1500 and serves the town's residential area, with the aim of encouraging car-owners to leave their vehicles at home and to travel into town by bus. Originally operated by route buses, the route has been provided on weekdays by a minibus in the Cancu Supreme fleet since 3rd October 2007.

FBY 720 is one of the most recent arrivals in the route bus fleet. A few low-floor buses have been imported from Britain since the scheme for funding new low-floor vehicles was abandoned. Formerly in the Stagecoach fleet it was seen on route 62 to Paceville coasting along the main esplanade by Balluta Bay.

ROUTES to MDINA, RABAT, DINGLI and MTARFA

Route 80 Valletta – Hamrun – Santa Venera – Mriehel – Balzan – Attard – Mdina – Rabat

The bus follows the same route to the Fleur de Lys roundabout as route 71, but then continues straight ahead along a broad dual carriageway which borders the Mriehel Industrial Estate. At a busy T-junction it turns right soon reaching a tree-lined road bordered by imposing and expensive houses on the edge of Attard. The centre of Attard is served by route 40. The built-up area soon gives way to an avenue through fields divided by rubble walls. The bluff on which Mdina and Rabat are built is now in view, over 150m above sea level, with Mdina's Baroque Cathedral prominent on the skyline. In the valley to the right across the fields the floodlights of Malta's National Stadium can be seen, close to the Ta'Qali Crafts Village.

The bus begins the steep ascent of the hill with the Saqqajja Bus Terminus on the right close to the summit. On outward journeys the bus bears left at the summit (avoiding Saqqajja) and passes along a high-walled road by the convent on the approach to Rabat. The bus makes a full clockwise circle of the town, along Triq had-Dingli, serving a suburb called Nigret, and turning right at another roundabout into Triq Gherixiem and then travelling uphill past the Roman Villa and the outskirts of Mdina and then to the terminus at Saqqajja.

Route 84 serves Mtarfa and then stops at Saqqajja terminus close to Rabat and the 'Silent City'. This Maltese (Zammit) bus with Debono bodywork is well over forty years old and is one of a small group of traditional route buses to have a passenger door. It was seen at Saqqajja.

Timings on route 80 from Valletta begin at 0550 and finish at 2130; from Rabat at 0530 until 2130. Journeys before 0930 (Mondays to Saturdays) serve Santa Rita in Rabat town centre.

Rabat and Mdina have been centres of population on Malta for centuries. The area is at the very heart of the region where Christianity was introduced to the island. In AD 60 St Paul is reputed to have slept for three months in a cave within a ditch beyond the walls of the old Roman town, after being shipwrecked on the coast. In the parish church of St Paul, first mentioned in documents as early as 1372, there is a huge painting of *The Shipwreck of St Paul* by Stefano Erardi. The Wignacourt College Museum displays the history of the College where the Knights who were hoping to become monks would reside. St Paul's Catacombs are an underground area containing numerous family burial plots of Punic, Roman, Jewish and Christian origins. These catacombs were cut out of live rock, and they form a labyrinth of tunnels and galleries. There is a chapel, and a main hall with agape (feasting) tables at both ends.

Mdina is known as the "Silent City", and was the island's capital city until 1568, when the Knights of St John built the new city of Valletta, following the Great Siege. Since that time Mdina has boasted another name — Citta Vecchia (The Old City) — to distinguish it from Valletta. Mdina traces its history back to Phoenician times (800 to 480 BC), when the city covered an area three times the size it does nowadays. The town's name reflects the Arabic "medina", meaning "walled town", from the era when the Arabs took Malta, redrew the city's boundaries and rebuilt the fortifications which still exist to this day. Mdina has a tiny population. Its extremely narrow streets prevent much road traffic passing through, as the city's residents are the only people allowed to bring their cars in. Mdina remains one of the world's finest examples of a medieval walled city which is still inhabited.

The Main Gate leads into St Publius Square, where the dungeons lie. The Palazzo Vilhena was the site of the mediaeval governing body of the island until the earthquake of 1693 devastated it. St Peter's Monastery, the Convent of the Sisters of St Benedict, is the home of about twenty nuns who are members of a strict enclosed order. The floor plan of St Paul's Cathedral is laid out in the form of a Latin cross. The cathedral has a Baroque façade with two ornate belfries and an octagonal dome which dominates the skyline for quite a distance around. The Bishop's Palace has been the residence of the Archbishops of Malta since 1722. Its design incorporated smaller windows than those of the cathedral, as it stands on the town's fortifications. Palazzo Falzon is the finest example of Norman architecture on the island and is now a privately owned museum which tells of the island's history over the past two centuries. Bastion Square offers the finest views of the island. Originally an artillery position, this area has ramparts overlooking the terraced fields and the island's central agricultural plain, Mtarfa, the Ta'Qali Crafts Village, and in the middle distance Mosta Dome. Local inhabitants claim that smoke rising from Mount Etna on Sicily (over eighty kilometres distant) can be seen on a very clear day.

Route 81 Valletta – Hamrun – Santa Venera – Mriehel – Balzan — Attard – Mdina – Rabat – Dingli

This route covers exactly the same roads as route 80 to Rabat. Where buses on route 80 bear right by St Dominic's Priory, vehicles on route 81 bear left, away from the urban area along fairly narrow and winding roads which lead to Verdala Palace and Buskett Gardens. At a crossroads close to the "Clapham Junction" cart ruts, buses turn right and right again and make the steady climb along Triq il-Buskett, past Savio College to the centre of Dingli. The buses make a complete circuit of the village's residential area, but departures from Valletta between 0930 and 1630 turn off the main roads through the town and head out along extremely rough lanes to Dingli Cliffs. At the terminus there is often a short break

Route 85 is a very infrequent route from Saqqajja terminus to Bahrija. Now numbered 85, this used to be merely an extension of route 80. FBY 787 is a Leyland Tiger Cub with a Scarnif Broadway body and it was photographed on the return journey from Bahrija.

before the return journey to Valletta is undertaken. Route 81 operates from Valletta from 0600 until 2200, mainly every thirty minutes.

Verdala Palace was designed in 1586 by Girolamo Cassar, the architect of St John's Co-Cathedral in Valletta, as a country villa and summer residence for the Grand Master Hugues Loubenx de Verdalle, a French cardinal. The palace looks very much like a traditional fortified mediaeval keep and today it is the official summer residence of the President of Malta.

The palace overlooks Buskett Gardens, which is Malta's most extensive wooded area. Originally the Gardens were developed as the hunting grounds for the palace where wild boar and deer were shot. The trees growing here include olives, oaks, pines and carobs; but few, if any, flowers grow here. The "Cart Ruts" are on a rocky plateau about four hundred metres from the Gardens; they are 30cm deep and 60cm wide and, although many instances are found in Europe, these Maltese "Ruts" are more numerous than anywhere else. Archaeologists still remain uncertain about their purpose and origin, but the Ruts are generally believed to date from Bronze Age (2,300 to 800 BC). Dingli Cliffs are the highest in Malta, rising sheer from the sea to a height of about 250 metres.

Route 82

This is a weekday variation of route 80 serving Ghar Barka estate just before reaching Nigret. There are six timings from Valletta, at 0605, 0710, 0750, 0820, 1620 and 1720.

Route 84 Valletta – Hamrun – Santa Venera – Mriehel – Balzan – Attard – Rabat – Mtarfa (circular)

This route serves Mtarfa, which is an expanding residential town outside Rabat. The town was formerly the site of British army barracks which were dominated by the David Bruce Military Hospital. The route covers the same ground as route 80, even as far as Nigret on the western edge of Rabat. Where the 80 turns right at the roundabout, 84 continues straight ahead, wends its way through a series of very narrow side-roads on the southern outskirts of Mtarfa, before making its way along the main road, past the school, and then turning right through a new housing development and circling its way back to the main road. Tracing its route back to the roundabout, the bus turns left and climbs back up the Saqqajja terminus at Mdina, as route 80 does too, passing the Roman Villa on the way.

Departures from Valletta are from 0720 until 1910 at half-hourly intervals, generally.

Route 85 Saqqajja Terminus — Bahrija

There are two departures to Bahrija, twelve hours apart, at 0515 and 1715 on weekdays, and a third at 1300 on Saturdays, which does not leave Bahrija on its return journey until 1330. Leaving Saqqajja, the bus skirts the walls of Mdina, passes the Roman Villa and descends to the roundabout on the approach to Mtarfa. The bus now sets out along roads not otherwise covered by bus routes, past Ghajn Qajjet, to Fiddien Bridge through the Ghemieri Valley along Triq il-Kuncizzjoni. A left turn and a steady climb bring the bus to Bahrija.

Route 65 Sliema Ferries – Paceville – San Gwann – Naxxar – Mosta – Ta' Qali Crafts Village – Rabat

This is another of the "Direct" services across the island. The bus takes the main route to Paceville and turns left at the traffic lights onto the Regional Road, passing under the tunnel at Ta'Giorni, taking the next exit for San Gwann. Travelling along the main road through the town, the bus is covering the same route here as route 41.

For the next kilometre or so the main bus route into Mosta is followed. After negotiating the lengthy one-way system in the town, the bus turns left along Triq il-Kbira and a quick series of left and right turns on the approach to Ta'Qali soon brings the bus close to the National Stadium where the Malta football team play its home matches. The roads hereabouts are built along what was the runway for the former World War Two airfield. The new US Embassy is being built here and the aircraft museum, Ta'Qali Crafts Village and glass factory are popular tourist attractions on this stretch of road. At the next T-junction, the bus turns right, joining the routes from Valletta near to the start of the ascent of Saqqajja Hill up to Rabat and the Saqqajja Bus terminus. Timings from Sliema begin at 0830 and operate primarily every half-hour until 1730. Return journeys from Rabat are from 0830 until 1830.

Route 86 Qawra – Mosta — Rabat – Mdina

This is another "Direct" service. From Qawra Bus terminus the bus follows the exact route of vehicles heading for Valletta to Mosta Dome. Here route 65 (above) is joined, and the bus similarly negotiates the one-way system and heads for Ta'Qali and Rabat along exactly the same route.

Timings are 0900 until 1800 from Qawra, and from 0930 until 1800 from Saqqajja terminus, basically every half-hour. Some of the early journeys from Saqqajja do not serve Mosta, instead travelling direct to Targa Gap.

ROUTES serving QORMI, ZEBBUG and SIGGIEWI

Route 88 Valletta – Hamrun – Qormi/St Sebastian – Zebbug

From Valletta, buses reach Hamrun which is a town large enough to boast two parishes, though with little to commend it to the tourist. The bus bears left by the convent, passes beneath the Regional Road at a large roundabout, where it continues straight on, past the Maltapost parcels depot and the Löwenbrau and Coca Cola factories. The route turns right at another roundabout and passes through the area of Qormi called St Sebastian, before joining the dual carriageway. The bus climbs the hill to Tal-Hlas with its tiny church in fields to the right. The original building of 1500 was razed to the ground in the island's earthquake of 1693. At the next roundabout the bus continues straight on, soon turning left for Zebbug along Triq il-Helsien which leads directly to the terminus in Triq Sciortino in front of the library and post office. Departures from Valletta are from 0600 until 2200 and from Zebbug from 0530 until 2130 at ten, fifteen or twenty minute intervals throughout the day.

Route 89 Valletta – Floriana – Hamrun – Qormi — Siggiewi

For much of its way to Siggiewi, this service follows route 88. At the roundabout, where route 88 bears right for Qormi, the bus on route 89 continues straight ahead along the broad Triq Manwel Dimech and then turns right onto the dual carriageway at its eastern end. Routes 88 and 89 divert again at the Tal-Hlas roundabout with route 89 bearing left along the main road to Siggiewi. The approach to the town centre involves negotiating the very narrow Triq Santa Margarit. The bus passes the parish church of St Nicholas, a magnificent example of Baroque architecture built between 1675 and 1693, and then it drops down a slight incline bearing right into the town square and the terminus in front of the chapel of St John the Baptist. Lying in the heart of the island's southern agricultural region, Siggiewi is one of the ten original parishes created in 1436 and nowadays has a population of about 6,000. Route 89 operates from 0600 until 2130 from Valletta, and from 0530 until 2030 from Siggiewi, every fifteen, twenty or thirty minutes depending on the time of day.

The chassis of the unique Reo Speedwagon is now over seventy years old while its bodywork is now fifty-five years old. Looking resplendent in the bright afternoon sunshine at the terminus of route 89 in Siggiewi, this wonderful vehicle, with registration DBY 368, was photographed back in March 2007.

King Long DBY 382 is seen descending the steep hill to Ghar Lapsi, which is an extension of the main route 89 from Siggiewi on Thursdays and Sundays during the high summer. Driver John Borg has not shown the 94 for this departure.

Route 94 Siggiewi – Ghar Lapsi

The other route in Siggiewi is really a seasonal extension of route 89, serving Ghar Lapsi on Thursdays and Sundays during July and August. Four departures on Thursdays are at 0800, 1100, 1500 and 2000. On Sundays there are more, at 0730, 1030, then hourly from 1400 until 1900 (but not at 1700).

The route heads into the rural area to the south of Siggiewi, past the Tal-Providenza Residential Home for the handicapped, which is a religious-based charity-funded organisation. The nearby Madonna tal-Providenza chapel is an octagonal building where mass is said just once a year, on the first Sunday in September. The bus then descends steeply to Ghar Lapsi with a fine view of Filfla Island out to sea. Ghar Lapsi, or Cave of the Ascension, attracts many local children to its small beach, along with scuba divers and local fishermen. The bus turns round in front of the café which is famous for its fish soup.

Route 91 Valletta – Hamrun – Qormi/St Sebastian – Qormi (St George) (circular)

This route serves another area of Qormi and so its route from Valletta mirrors route 88 as far as Triq il-Vittorja in Qormi. From this point there are variations of the route dependent on the time of day or day of the week. All services terminate by the parish church of St George in Triq Kardinal Xibberas. On the return journey buses thread their way along Triq il-Kbira and Tri il-Ghaqda to rejoin the outward route near the Post Office, and so back to Valletta. Departures from Valletta are between 0600 and 2100, and from Qormi from 0530 until 2000 every twenty or thirty minutes.

HOSPITAL and UNIVERSITY ROUTES

Route 75 Valletta – Hamrun – St Luke's Hospital – Mater Dei Hospital

This long-standing route pre-dates the University routes described below. The bus reaches Hamrun, and soon turns right by the main police station into Triq Ferrovija; keeping straight on, it passes through the township of Gwardamanga and reaches St Luke's Hospital. This was the long-standing terminus of the route, but with the opening of the Mater Dei Hospital, the bus now plunges down Triq Bordin, soon reaching the main road near Msida roundabout. Then, via the short climb to the Regional Road to Tal-Qroqq roundabout, and a short section of the Birkirkara Bypass, the bus bears left along access roads to the terminus in front of the Mater Dei Hospital. On the return journey the bus continues round the edge of the hospital, with a bus-stop at the Maternity Unit, via the University terminus, back to Msida roundabout, turning left and climbing up to St Luke's again, and then back on the traditional route to Valletta. The route operates every fifteen or twenty minutes throughout the day.

An unusual feature of this route number is that some of the principal routes on the island include their own route 75 departure from their outer terminus at about 1500 each day, so as to arrive at the hospitals in time for the start of hospital visiting at 1530.

On Monday 11th October 2004, six new routes were introduced to provide direct access by bus from six areas of the island to the new University at Tal-Qroqq. For the first year or so the routes extended to Paceville and Sliema Ferries, but patronage proved to be poor and so the Sliema section was withdrawn on all routes in early October 2005. At the same time, the routes were revised to serve the Regional College in Msida. Until July 2007 all the routes operated only on week-days during term-time.

On Sunday 1st July 2007 the official opening of the Mater Dei Hospital took place, routes now ran daily throughout the year with extended hours of operation with four more routes added to the group. At the same time Route 58 from Qawra to Valletta was enhanced and re-routed via the Hospital and University on inward journeys. With the provision of the new routes (below), these special journeys have been far fewer than previously, since 3rd December, 2007.

Buses now park at designated bus-stops and timings have been reviewed in the interim so as suit students' lecture-times better and hospital visiting hours and shift-workers' hours. The mass exodus of all the buses on these routes at ten-past the hour is recommended to all bus photographers!

Route 110 Birzebbugia – Marsaxlokk – Zejtun – Tarxien – Paola – Junior College – University

Route 300 Senglea – Vittoriosa – Zabbar – Fgura – Paola – Junior College – University

Route 350 Mqabba – Qrendi – Zurrieq – Safi – Kirkop – Luqa – Marsa – Junior College – University

Route 800 (Airport) – Gudja – Ghaxaq – Bir-id-Deheb – St Lucija – Paola – Marsa – Junior College – University

Route 810 Dingli – Rabat – Attard – San Anton Gardens – Junior College – University

Route 890 Siggiewi – Zebbug – Qormi – Hamrun – Junior College – University

These six routes depart from the outer termini at 0645, 0745, then hourly between 0900 and 1900. Departures from the University are hourly between 0710 and 1910, then 2030.

Route 450 Cirkewwa – Mellieha – St Paul's Bay – Mosta – Lija – Birkirkara (Ganu Street) – Birkirkara Bypass – Hospital – University

King Long EBY 548 is parked in front of the University of Malta at Tal Qroqq, waiting to depart for Siggiewi on route 890.

Departures from Cirkewwa are at 0630, then hourly from 0700 until 1900; from the University, hourly from 0710 until 1910

Route 560 Gharghur – Naxxar – Mosta – Lija – Birkirkara Bypass – Hospital – University

Departures from Gharghur are hourly from 0600 until 0900, then from Naxxar 1000 until 1900; from the University, hourly from 0710 till 1910, and then 2030.

Route 580 Qawra – St Paul's Bay – Mosta – Lija – Birkirkara (Naxxar Road) – Valley Road – Birkirkara – Hospital – University

Hourly from Qawra between 0700 and 1700; from the University, hourly from 0810 until 1810.

Route 675 St Andrew's – St Julian's – Gzira – Tal Qroqq roundabout – Birkirkara Bypass – Hospital — University

Departures at 0610, 0705, then hourly from 0800 until 1900; from the University, hourly from 0710 until 1910.

Route 169 Valletta – Msida – Swatar – Hospital – University – Msida – Valletta (circular)

This route, just like route 75, pre-dates the University services. After reaching Msida, the bus heads towards Birkirkara and turns right onto the Regional Road, serving a bus-stop on the slip-road near Triq it Torri. It then makes a full circuit of Swatar and then serves the Hospital and University grounds en route back to Valletta.

Week-day timings are hourly between 0640 and 1840; Saturdays hourly until 1240, then 1500; Sundays, 0900, 1000 and 1500.

ROUTES 98 and 198 around VALLETTA

These routes are the only ones to serve Valletta itself. The very nature of the capital city with its narrow, hilly streets and the pedestrianisation of the principal shopping thoroughfares preclude the possibility of route buses even entering the city's heart. So route 198 makes an anti-clockwise circuit of the main roads which surround the city along the harbours' perimeters. Valletta must therefore be able to claim to be the capital city with the fewest bus routes actually operating through its streets.

From the Bus Terminus, the bus serves Floriana and then passes the Park and Ride car-parks at Blata l-Bajda before travelling along the quay edge where cruise liners are moored. Of course, passengers from these liners travel on this route to reach Valletta for a day's visit. The bus passes under Fort Lascaris and then begins the circuit of the main road around the Valletta peninsula, passing the Siege bell which is rung at midday every day, the Mediterranean Conference Centre and its "Malta Experience" which is an audio-visual history of the city and island, the War Museum and Fort St Elmo and the outer defence walls with their bastions built at focal points to protect the city. Across Marsamxett Harbour, Tigne and Sliema can be seen. The bus soon reaches the Phoenicia Hotel and turns left soon reaching the Castle Gate terminus.

Route 198 operates every half-hour on weekdays between 0930 and 1800, on Saturdays between 0730 and 1800 and on Sundays between 0830 and 1800. Route 98 is the clockwise variation of the route which operates on weekdays only in the early morning from 0630 until 0900.

BMC Falcon FBY 755 is seen operating the Valletta service, route 198 around the peninsula. The driver stopped for a few minutes for this photo shot along Triq Marsamxett near the German Curtain, in March 2009.

UNSCHEDULED BUSES ON MALTA

The Unscheduled Buses on Malta are the tour coaches of the island. Indeed, for most tourists arriving by plane at Malta International Airport, a tour coach is the first PSV that they are likely to see and use. Totalling almost 150, these coaches provide the journeys between the holiday resorts and the airport for holidaymakers, and, of course, the daily tours around the island, and connections from hotels to the Gozo Ferry at Cirkewwa for day-trippers to Malta's smaller neighbour.

Until the early 1980s coaches on Malta bore an all-over cream livery. Many, if not all of them, carried locally built bodywork. The livery was then changed to blue and cream. As newer coaches arrived on the island, many of the cream-coloured coaches were cascaded down to operate as route buses on the island. The foregoing route bus fleet-list clearly indicates this change in use vehicle by vehicle. During 1986 and 1987 the main influx of coaches from the United Kingdom began. They too were painted in the blue and cream livery, and a few of them can still be seen on the island nowadays. Many of these early imports were Plaxton-bodied coaches without opening side windows. They soon became unpopular with passengers, as they offered little or no accessible ventilation, especially to offset the high average summer temperatures on the island. Consequently individual owners installed air-conditioning to alleviate the problems caused by the heat on board these coaches.

The vast majority of the coaches had always remained "anonymous" – that is, they carried no indication of their owner's name or a fleet name. But a gradual change took place from the early years of the 1990s, as more and more of the coaches — still in the blue and cream livery — appeared with their operator's name proudly displayed. In 1995 a de-restriction in coach liveries was introduced which allowed coach operators to devise their own, as the authorities wished to make the island's coaching scene brighter.

Paramount Garage's Volvo B10M with Irizar Century bodywork, ACY 974 is seen in the main parking area at Castle Gate in Valletta in September 2009. Its predominantly purple livery is in contrast to many other coaches in this fleet.

LCY 949 is an ERF E10 Trailblazer with Unicar bodywork and is part of the Zarb Coaches fleet. It had only recently received this contact livery when seen near the radar dome on Dingli Cliffs in September 2009.

Some of the larger coaching firms were very quick to seize the opportunity to individualise their fleet. Garden of Eden was the first company to set the trend with the artistic paintings of local buildings adorning the rear panels. Tourism on Malta had obviously grown over the years, and the PTA recognised the need for a larger coaching fleet to cope with the ever-increasing demands of the tourist companies. So these changes coincided with the importation of the Plaxton Paramount coaches from Britain.

The KopTaCo co-operative was formed in 1997 by some of the smaller coach companies and its sixteen members now benefit from the economies of scale which the operation has brought. A corporate livery has developed within the fleet.

This co-operative has therefore seen the disappearance of some of the smaller companies which featured in the first edition of this *Bus Handbook*, such as Morin Coaches and Ventura.

Nowadays the age profile of the coaches ranges from some AEC Reliances with Plaxton bodywork imported second-hand from Britain which are nearly 35 years old through the more modern coaches with bodywork built by Ghabbour in Egypt and Irizar in Spain to the King Long coach which arrived in 2001. The last decade has seen the continued updating of the Unscheduled fleet with more modern, second-hand coaches on Volvo and Scania chassis being imported. Added to which, rebuilding of coaches is common by their owners and such details are referred to in the fleet-list below. In 2009 two demonstrator Temsa coaches with the new xPY registration entered service with Cancu Supreme.

UNSCHEDULED BUSES (TOUR COACHES)

JCY 850	Dennis Javelin 11SDA1906	Plaxton Paramount 3200	C53F	1/89	6/93	Leone Grech, Mosta Paramount Garage
KCY 851	Ford R1114	Caetano Alpha	C53F	11/79	4/86	Victor Muscat, Naxxar KopTaCo
JCY 852	Dennis Javelin 11SDA1906	Plaxton Paramount 3200 II	C53F	1/90	5/93	Leone Grech, Mosta Paramount Garage
KCY 853	Bedford YNT	Plaxton Paramount II	C53F	8/85	by 10/91	Dalli Bros., Gzira Welcome Garage
ACY 854	Volvo B10M	Platon Première	C53F	3/98	3/05	Nazzareno Abela, Zetjun Cancu Supreme
KCY 855	Bedford YMT	Duple Dominant II	C53F	1/80	by 8/86	Alphonse Abela, Birzebbugia Peppin Garage
LCY 856	Bedford YNT	Plaxton Paramount 3200 II	C53F	5/85	by 12/91	John Galea, Balzon Arthur and John's
LCY 857	Bedford YMT	Van Hool McArdle	C53F	5/78	by 1/87	John Galea, Balzon Arthur and John's
LCY 858	Scania K113CRB	Irizar Century	C53F	6/95	8/04	John Galea, Balzon Arthur and John's
ACY 859	Bedford YMT	Plaxton Supreme III	C53F	6/77	by 10/86	Joseph Mary Farrugia, Mosta KopTaCo
LCY 860	Dennis Javelin GX	Plaxton Excalibur	C53F	1997	c6/00	Leone Grech, Mosta Paramount Garage
	(This vehicle carries its Plaxton demonstration livery used prior to its importation to Malta)					
BCY 861	Optimal	Scarnif	C53F	10/01	10/01	Nazzareno Abela, Zejtun Cancu Supreme
BCY 862	Volvo B10M-62	Plaxton Première 350	C53F	6/96	by11/02	Nazzareno Abela, Zejtun Cancu Supreme
ACY 863	Leyland Tiger TRCL10/3ARZM	Plaxton Paramount 3500 III	C53F	1/91	by 4/98	Joseph Spiteri, Ghaxaq Garden of Eden
KCY 864	Bedford YMT	Plaxton Supreme IV Express	C53F	1/80	by 8/66	Charles Dalli, Gzira Welcome Garage
JCY 865	Volvo B10M-62	Plaxton Première 350	C53F	5/96	by 10/02	Charles Dalli, Gzira Welcome Garage
ACY 866	Volvo B10M-62	Plaxton Première 350	C53F	5/96	3/01	Anthony Grima, Mellieha KopTaCo
LCY 867	Bedford YMT	Plaxton Supreme III Express	C53F	3/77	by 10/86	Charles Gauchi, Mosta KopTaCo
JCY 868	Bedford YNT	Plaxton Paramount 3200 III	C53F	3/87	5/89	Leone Grech, Mosta Paramount Garage
JCY 869	Bedford YNT	Plaxton Paramount 3200	C54F	6/84	by 7/87	Leone Grech, Mosta Paramount Garage
KCY 870	Bedford YMT	Duple Dominant II	C53F	4/79	4/88	Carmel Zammit, Mosta KopTaCo

JCY 850, Y-0850, F666PAY   Snowdon, Easington Colliery
KCY 851, Y-0851, CNY333V   Eynon, Trimsaran: Thomas, Clydach Vale
JCY 852, Y-0852, G966WNR   Dunnet, Keiss
KCY 853, Y-0853, C350FBO   Avondale, Rothwell
ACY 854, R412FWT   Fowler, Holbeach Drove: Wallace Arnold, Leeds
KCY 855, Y-0855, EHE225V   Anderson, Westerhope: National Travel East, 225
LCY 856, Y-0856, B624DDW   GP Travel, London
LCY 857, HVO20V   Luxicoaches, Borrowash
LCY 858, M276ROS, HSK857, M499XWF   Youngs, Ross: Buddens, Romsey
ACY 859, Y-0859, SBH107R   K Line, Huddersfield
LCY 860   Demonstrator in Hong Kong *(Stored from 4/99 to c6/00)*
BCY 861   New
BCY 862, N242HWX   Wallace Arnold, Leeds
ACY 863, H381TNG   Elsey, Gosberton
KCY 864, Y-0864, KBH856V   Clarkes of London, SE 20
JCY 865, N212HWX, 96-D-25736   Cityliner, Port Glasgow: Bus Eireann, VP13
ACY 866, N758BNU   Dunn–Line, Nottingham
LCY 867, Y-0867, PNK150R   Scotland & Bates, Appledore
JCY 868, Y-0868, D371KDB   Bullock, Cheadle
JCY 869, Y-0869, A111MAC   Harry Shaw, Coventry
KCY 870, Y-0870, BRY63T   Mundy, Orpington

KCY 888 is one of the few coaches on Malta with a Caetano Alpha body. Its plain white livery does not identify it as part of the KopTaCo co-operative. This view of the coach was taken at Cirkewwa.

CCY 871	Bedford YMT	Duple Dominant II	C53F	5/80	by 3/87	Saviour Borg, Msida KopTaCo
KCY 872	Bedford YMT	Duple Dominant II	C53F	8/77	by 3/86	Paul Muscat, Naxxar KopTaCo
KCY 873	Bedford YMT	Duple Dominant II	C53F	3/79	by 10/87	Emanuel Vella, Qormi KopTaCo
LCY 874	Scania L94 IB	Irizar Century	C50FT	1/99	5/08	Nazzareno Abela, Zejtun Cancu Supreme
KCY 875	Bedford YMT	Duple Dominant II	C53F	4/78	by 3/86	Joseph Saliba, Zurrieq Nimrod Coaches
BCY 876	Bedford YMT	Scarnif	C53F	8/78	by 5/87	Mario Sultana, Gzira Sultana
KCY 877	Volvo B10M	Plaxton Première 350	C53F	2/94	by 11/02	Mario Sultana, Gzira Sultana
ACY 878	Bedford YMT	Caetano Alpha	C53F	10/79	by 8/86	Kevin Muscat, Naxxar KopTaCo
KCY 879	AEC Reliance 6U3ZR	Plaxton Supreme IV	C53F	1/79	by 7/87	Paul Vella, Mosta Silver Star
CCY 880	Bedford YMT	Plaxton Supreme IV	C53F	4/80	3/86	Charles Borg, Qormi KopTaCo
BCY 881	Scania K124	Van Hool T9	C49Ft	3/00	12/06	Nazzareno Abela, Zejtun Cancu Supreme
JCY 882	Volvo B10M-60	Plaxton Première 350	C53F	5/92	7/01	Emmanuel Zarb, Birkirkara Zarb Coaches

CCY 871, Y-0871, GPA631V   Arrowline Coaches, West Drayton
KCY 872, Y-0872, UTV813S   Rodgers, Langwith
KCY 873, Y-0873, UWH694T   Bailey, Kirkby
LCY 874, S364SET, 99-D-8473   Dockyard Coaches, London; Bus Eireann S136.
KCY 875, Y-0875, YHJ169S   Poulson, Copford
BCY 876, Y-0876, YPB822T   Grimshaw, Burnley
KCY 877, L3TCC   Johnson, Hodthorpe: Travellers, Hounslow
ACY 878, Y-0878, KUB962V   Aris, Long Harborough:Murgatroyd, Thurnscoe
KCY 879, Y-0879, BGY583T   New Enterprise, Tonbridge
CCY 880, Y-0880, LVS435V   Reliance, Newbury, 170: Pearl Line Coaches, London SW6
BCY 881, W222KDY   Rambler, Hastings
JCY 882, OIL5640, J716CWT   Associated Motorways, Harlow

LCY 885 is a Leyland Tiger with Plaxton Paramount 3200 bodywork with modifications made in-house at Zarb. The coach is seen awaiting its passengers from the Dolmen Hotel in Qawra for a day excursion.

JCY 883	AEC Reliance	Plaxton Supreme IV (1979)	C53F	5/67	4/87	Emmanuel Zarb, Birkirkara Zarb Coaches
KCY 884	Optimal Leyland	Scarnif	C53F	9/01	2001	Emmanuel Zarb, San Gwann Zarb Coaches
LCY 885	Leyland Tiger TRCTL11/2R	Plaxton Paramount 3200/Zarb	C53F	1/84	by 10/91	Joseph Zarb, Naxxar Zarb Coaches
ACY 886	Bedford YMT	Duple Dominant II	C53F	1/80	3/87	Jane Sammut, Luqa KopTaCo
ACY 887	Volvo B10M	Berkhof Axial	C49Ft	3/98	12/06	Emmanuel Zarb, San Gwann Zarb Coaches
KCY 888	Ford R1114	Caetano Alpha	C53F	11/80	4/88	Paul Muscat, Naxxar KopTaCo
KCY 889	Dennis Javelin 11SDA1906	Plaxton Paramount 3200 III	C53F	3/88	by 4/94	Nazzareno Abela, Zejtun Cancu Supreme
KCY 890	Bedford YNT	Plaxton Paramount 3200 II	C53F	2/87	6/91	Leone Grech, Mosta Paramount Garage
KCY 891	Ford R1114	Plaxton Supreme IV	C53F	2/79	5/87	Anthony Grima, Mellieha KopTaCo
LCY 892	Bedford YMT	Plaxton Supreme IV	C53F	3/79	by 10/87	Salvino Farrugia, Paola KopTaCo
LCY 893	Scania L94	Van Hool Alizée T9	C53F	4/99	3/09	Godwin Farrugia, Mosta KopTaCo
ACY 894	Volvo B10M-60	Plaxton Paramount 3200	C53F	5/91	11/01	Joseph Saliba, Zurrieq Nimrod Coaches

JCY 883, Y-0883, CPM520T, HOD39E   Thomas, West Ewell, (*This vehicle was rebuilt from AEC Reliance 2U3RA Duple Northern in 1978 and rebodied in 1979*)

KCY 884   New
LCY 885, Y-0885, A633XFM, 614BWU, A423LRJ   Barry Cooper, Stockton Heath
ACY 886, Y-0886, FUJ920V   Whittle, Highley, 20
ACY 887, R392NUF, TDY388, R222VDY   Rambler, Hastings
KCY 888, Y-0888, NAY428W   Cedric, Wivenhoe
KCY 889, Y-0889, E270AJL   Hornsby, Ashby
KCY 890, Y-0890, 9424RU, D370KDB   County, Brentwood
KCY 891, Y-0891, WHC538T   WHM, Little Walthorn
LCY 892, Y-0892, CEL104T   Winson, Loughborough
LCY 893, T90ASH   Ashton Coaches, St Helens
ACY 894, H112OON   Zarb Coaches, Londoners, Nunhead

It is rare for an imported coach on Malta not to be repainted into the fleet livery of its new owner, but BCY 902, a Volvo B10M with Van Hool Alizée bodywork is an exception. It retains the green livery of Rambler of Hastings, but with Cancu Supreme fleet names added. The "Supreme of Zejtun" is another unusual logo for coaches in Reno Abela's fleet. The coach is seen near Ta'Qali.

LCY 895	Bedford YMT	Duple Dominant II	C53F	5/78	by 3/86	Joseph Schembri, Siggiewi Schembri Coaches
LCY 896	Volvo B10M-62	Plaxton Première 350	C53F	2/95	11/02	Mario Sultana, Gzira Sultana
LCY 897	Bedford YNT	Plaxton Paramount 3200 II	C53F	11/85	by 10/91	Mario Sultana, Gzira Sultana
KCY 898	Bedford YMT	Plaxton Supreme IV	C53F	11/78	2/87	Joseph Zarb, San Pawl il-Bihar individual owner
JCY 899	Bedford YMT	Duple Dominant II	C53F	4/78	by 8/86	Joseph Calleja, St Julian's Garden of Eden
LCY 900	Ford R1114	Caetano Alpha	C53F	1/80	by 8/86	Andrew Spiteri, Luqa Century Coaches
BCY 901	Ford R1114	Scarnif (1998)	C53F	4/79	by 10/86	Salvu Abela, Zejtun Cancu Supreme
BCY 902	Volvo B10M	Van Hool Alizée	C53FT	1/96	3/05	Catherine Abela, Zejtun Cancu Supreme
ACY 903	AEC Reliance 6U3ZR	Duple Dominant II	C53F	8/79	by 1/87	Andrew Abela, Zejtun Cancu Supreme
CCY 904	Bedford YMT	Plaxton Supreme Express	C53F	4/78	by 10/87	Joseph Abela, Birzebbugia Swallow Garage
JCY 905	Leyland Royal Tiger PSU1/13	?: Daina (1974): Paramount Garages (1995)	C41F	5/53	7/75	Leone Grech, Mosta Paramount Garage
ACY 906	Bedford YMT	Duple Dominant II	C53F	1/79	by 10/87	Anthony Zahra, San Gwann Zarb Coaches

LCY 895, Y-0895, UUT786S — Arrowline, Hayes
LCY 896, M870EEA — Goode, West Bromwich
LCY 897, Y-0897, C432HHL — Wainfleet, Nuneaton
KCY 898, Y-0898, CRW517T — Bailey, Kirkby
JCY 899, Y-0899, VNT1S — Arleen, Peasedown
LCY 900, Y-0900, HAY800V — Globe, Barnsley: Euro Academy, Croydon
BCY 901, Y-0901, YCF966T — Reliance, Newbury, 160 (*This vehicle was rebodied in 1998*)
BCY 902, N625PYJ, N222EDY — Rambler, Hastings
ACY 903, Y-0903, LBD929V — Vaughan, Salford: Brittain, Northampton
CCY 904, Y-0904, WRY 90S — Don, Dunmow
JCY 905, Y-0905, Y-1605, A-1982, 1982, GUH462 — Skey, Sywell
ACY 906, Y-0906, AUJ734T — New Enterprise, Tonbridge

Another example of a coach retaining the basic livery of its English heritage is ACY 914, Anthony Grima's Volvo B10M with Van Hool bodywork. In this example KopTaCo fleet names have been applied to the Shearings livery. The coach is parked close to St James's Bastion in the lay-over area at Castle Gate bus terminus in Valletta.

LCY 907	King Long XMQ 6113	King Long	C49F	7/01	7/01	Nazzareno Abela, Zejtun Cancu Supreme
ACY 908	Volvo B10M-62	Plaxton Excalibur	C50F	4/94	1/02	Andrew Abela, Zejtun Cancu Supreme
ACY 909	Scania K93CRB	Plaxton Paramount 3200 III	C53F	4/91	by 4/95	Jane Sammut, Luqa KopTaCo
LCY 910	Volvo B10M-61	Plaxton Paramount 3500 II	C53F	3/85	by 10/91	John Galea, Balzan Arthur and John's
ACY 911	Volvo B10M-60	Plaxton Première 350	C53F	3/93	2/02	Mario Abela, Zejtun Cancu Supreme
ACY 912	Volvo B10M-62	Plaxton Première 350	C53F	3/95	c4/00	Nazzareno Abela, Zejtun Cancu Supreme
BCY 913	Bedford YNT	Plaxton Paramount 3200 II	C53F	6/87	by 4/93	Nazzareno Abela, Zejtun Cancu Supreme
ACY 914	Volvo B10M	Van Hool T9	C53F	3/00	5/07	Anthony Grima, Mellieha KopTaCo
ACY 915	AEC Reliance	Plaxton Supreme IV (1981)	C53F	7/66	by 3/90	Anthony Grima, Mellieha KopTaCo
BCY 916	Volvo B10M-62	Plaxton Première 350	C53F	4/97	4/03	Mario Abela, Zejtun KopTaCo
ACY 917	Volvo B10M-62	Plaxton Première 350	C53F	3/95	5/01	Nazzareno Abela, Zejtun Cancu Supreme

LCY 907   New
ACY 908, L930NWW   Smith, Market Harborough
ACY 909, Y-2501, H926DRJ   Shearings, 926
LCY 910, Y-0910, B367RHC, NDY820, B191XJD   Rambler, Hastings
ACY 911, K624FEC, K3JFS   Fishwick, Leyland C4
ACY 912, M39KAX   Bluebird Buses
BCY 913, Y-0913, D384CFR, BIB5428, D439GAD   Bibby, Ingleton
ACY 914, W202JBN   Shearings, 202
ACY 915, Y-0915, OPD789W, KHM5D   Thomas, West Ewell (*This vehicle was rebuilt from AEC Reliance 3U2RA Willowbrook and rebodied in 1981*)
BCY 916, P20TGM   Tellings-Golden Miller, Heathrow 1151
ACY 917, M171EYG, 8980WA, M103UWY   Wallace Arnold, Leeds

CCY 920 is a Ford R1114 with Plaxton Supreme IV bodywork. It is one of Paul Debono's coaches in his Bonu Coaches fleet. It is seen in this view parked next to the Triton Fountain at the Castle Gate bus terminus in Valletta in June 2009. It is interesting to compare the current less flamboyant livery of this coach with the one featured on page 74 of the first edition of this *Bus Handbook*. *T W W Knowles*

ACY 918	Bedford YMT	Scarnif (2009)	C53F	4/77	by 3/86	Mario Abela, Zejtun Cancu Supreme
KCY 919	Ford R1114	Duple Dominant II	C53F	9/79	2/86	Anthony Grima, Mellieha KopTaCo
CCY 920	Ford R1114	Plaxton Supreme IV	C53F	4/79	2/86	Paul Debono, San Gwann Bonu Coaches
LCY 921	Bedford YMT	Plaxton Supreme IV	C53F	5/80	5/87	Joseph Spiteri, Ghaxaq Garden of Eden
LCY 922	Bedford YMT	Duple Dominant II	C53F	4/80	by 3/86	Joseph Mifsud, Gzira KopTaCo
LCY 923	Optimal	Scarnif	C53F	4/98	4/98	Anthony Zahra, San Gwann Zarb Coaches
BCY 924	Volvo B10M	Plaxton Première 350	C53F	4/98	5/03	Nazzareno Abela, Zejtun Cancu Supreme
JCY 925	Dennis Javelin	Plaxton Paramount 3200 III	C53F	1/88	5/93	Leone Grech, Mosta Paramount Garage
xCY 926	(This registration has never been issued)					
xCY 927	(This registration has never been issued)					
xCY 928	(This registration has never been issued)					
CCY 929	Dennis Javelin	Berkhof Excellence 1000	C53F	8/91	1/95	Joseph Spiteri, Ghaxaq Garden of Eden

ACY 918, Y-0918, SNR632R   Brockbank, Staveley
KCY 919, Y-0919, CYH801V   Cowie: Grey Green, London N16
CCY 920, Y-0920, AUJ523T   Wilkinson, Hebburn
LCY 921, Y-0921, UHJ351V, 5919RU, MMJ473V   County, Brentwood: AC Coaches, London SE15
LCY 922, Y-0922, GPA630V   Arrowline, Coaches, West Drayton
LCY 923   New
BCY 924, R435FWT   Vale of Llangollen, Cefn Mawr
JCY 925, Y-0925, E842EUT   Snowdon, Easington Colliery
xCY 926
xCY 927
xCY 928
CCY 929 Y-0929, J10BCK   Sanders, Holt

BCY 930	Dennis Javelin	Berkhof Excellence 1000	C53F	5/91	1/95	Joseph Spiteri, Ghaxaq Garden of Eden
BCY 931	Leyland Tiger TRCL10/3RZM	Plaxton 321	C53F	5/91	1/95	Carmelo Abela, Zejtun KopTaCo
ACY 932	Leyland Tiger TRCL10/3RZM	Plaxton Paramount 3200 III	C53F	1/92	1/95	Emanuel Zarb, Birkirkara Zarb Coaches
BCY 932	*This registration is no longer allocated. The following coach is its replacement.*					
GPY 014	Scania K124	Irizar Century	C49Ft	12/01	8/09	Anthony Grima, Gzira KopTaCo
BCY 933	Leyland Tiger TRCL10/3ARZM	Plaxton Paramount 3200 III	C53F	10/91	1/95	Carmel Zarb, Birkirkara Zarb Coaches
BCY 934	Leyland Tiger TRCL10/3ARZM	Plaxton 321	C53F	8/91	1/95	Nazzareno Abela, Zejtun Cancu Coaches (SMS Tours contract livery)
LCY 935	Leyland Tiger TRCL10/3ARZM	Plaxton Paramount 3200 III	C53F	4/92	3/95	Esther Grima, Mellieha KopTaCo
JCY 936	Leyland Tiger TRCL10/3ARZM	Plaxton 321	C53F	8/91	1/95	Victor Muscat, Naxxar KopTaCo
JCY 937	Leyland Tiger TRCL10/3ARZM	Plaxton 321	C53F	8/91	1/95	Paul Muscat, Naxxar KopTaCo
BCY 938	Leyland Tiger TRCL10/3ARZM	Plaxton Paramount 3500 III	C53F	2/91	1/95	Salvu Abela, Zejtun Cancu Supreme
KCY 939	Volvo B10M-60	Plaxton Paramount 3200 III	C53F	8/92	1/95	Anthony Grima, Mellieha KopTaCo
LCY 940	Leyland Tiger TRCL10/3ARZM	Plaxton Paramount 3500 III	C53F	2/92	1/95	Joseph Abela, Ghaxaq Cancu Supreme
KCY 941	Dennis Javelin	Plaxton Première 320	C53F	4/93	1/95	Nazzareno Abela, Zejtun Cancu Supreme
ACY 942	Leyland Tiger TRCL10/3ARZM	Plaxton 321	C53F	2/91	1/95	Joseph Farrugia, Mosta KopTaCo
KCY 943	Leyland Tiger TRCL10/3ARZM	Plaxton 321	C53F	2/91	1/95	Salvinu Farrugia, Paola KopTaCo
CCY 944	Leyland Tiger TRBL10/3ARZA	Plaxton Paramount 3200 III	C53F	5/91	4/95	Emanuel Zarb, Birkirkara Zarb Coaches
BCY 945	Leyland Tiger TR2R62C21Z6/8	Plaxton Paramount 3200 III	C53F	6/91	1/95	Leone Grech, Mosta Paramount Garage
LCY 946	Leyland Tiger TRCL10/3ARZM	Plaxton Paramount 3200 III	C53F	12/91	1/95	Gabriel Vassallo, Rabat KopTaCo
CCY 947	Leyland Tiger TR2R62C21Z6	Plaxton Paramount 3200 III	C53F	5/91	3/95	Nazzareno Abela, Zejtun Cancu Supreme
LCY 948	ERF E10 Trailblazer	Unicar	C53F	11/95	11/95	Joseph Spiteri, Ghaxaq Garden of Eden
LCY 949	ERF E10 Trailblazer	Unicar	C53F	11/95	11/95	Emanuel Zarb, Birkirkara Zarb Coaches
ACY 950	Volvo B10M-60	Plaxton Première 350	C53F	4/94	2/95	Catherine Abela, Zejtun Cancu Supreme

BCY 930, Y-0930, H10GSM — Sanders, Holt
BCY 931, Y-0931, H2RAD — Dunn-Line, Nottingham
ACY 932, Y-0932, J385ARR, J8DLT — Dunn-Line, Nottingham
(BCY 932)
GPY 014, 01-KK-2263 — Whorton, Crossedoney, Eire
BCY 933, J7DLT — Dunn-Line, Nottingham
BCY 934, Y-0934, J328PDE, A11WLS, J47SNY — Silcox, Pembroke Dock
LCY 935, Y-0935, J64KMR — Ellison, Ashton Keynes
JCY 936, Y-0936, J53SNY — Evans, Tregaron
JCY 937, Y-0937, J56SNY — Evans, Tregaron
BCY 938, Y-0938, H210AKH — Porteous, Anlaby
KCY 939, Y-0939, K6RAD — Dunn-Line, Nottingham
LCY 940, Y-0940, J780HAT — Porteous, Anlaby
KCY 941, Y-0941, K365HBE, 8227RH, K119GFU — Hornsby, Ashby
ACY 942, Y-0942, H331NNY — Thomas Rhondda
KCY 943, Y-0943. H332NNY — Thomas Rhondda
CCY 944, Y-0944, H263GRY — Reliance, Gravesend
BCY 945, Y-0945, H266GRY — Reliance, Gravesend
LCY 946, Y-0946, J260MFP — Pullman, Crofty
CCY 947, Y-0947, H265GRY — Reliance, Gravesend
LCY 948, Y-0948 — New
LCY 949, Y-0949 — New
ACY 950, Y-0950, L981ORB — Dunn-Line, Nottingham

LCY 951	Scania K93CRB	Plaxton Paramount 3200 III	C53F	2/91	by 5/95	Josephine Cassar, Qormi KopTaCo
LCY 952	Scania K93CRB	Plaxton Paramount 3200 III	C53F	4/91	by 5/95	Joseph Borg, Qormi KopTaCo
KCY 953	Scania K93CRB	Plaxton Paramount 3200 III	C53F	4/91	by 4/95	Joseph Saliba, Zurrieq Nimrod Coaches
JCY 954	Volvo B10M-60	Plaxton Paramount 3500 III	C53F	6/90	by 3/95	Mario Sultana, Gzira Sultana
JCY 955	Volvo B10M-60	Plaxton Première 350	C53F	5/92	by 5/95	Maria Sultana, Gzira Sultana
CCY 956	Volvo B10M-60	Plaxton Paramount 3500 III	C54F	3/91	by 4/95	Emanuel Zarb, Birkirkara Zarb Coaches
JCY 957	Volvo B10M-60	Plaxton Paramount 3500 III	C53F	3/91	by 4/95	Paul Debono, San Gwann Bonu Coaches
ACY 958	Dennis Javelin	Plaxton Première 320	C53F	4/93	2/95	Leone Grech, Mosta Paramount Garage
BCY 959	Scania K93CRB	Ghabbour	C55F	11/95	11/95	Emanuel Vella, Mgarr KopTaCo (Meeting Point contract livery)
ACY 960	Scania K93CRB	Ghabbour	C55F	11/95	11/95	Andrew Spiteri, Luqa Cancu Supreme
LCY 961	Scania K93CRB	Ghabbour	C55F	11/95	11/95	Joseph Schembri, Siggiewi Schembri Coaches
KCY 962	*This vehicle was destroyed by fire in 1/09*					Andrew Pace, Gzira KopTaCo
BCY 963	Scania K93CRB	Ghabbour	C55F	11/95	11/95	Joseph Zarb, San Pawl Il-Bahar KopTaCo
ACY 964	MAN 11.190 HOCL-R	Caetano Algarve 2	C35F	c11/95	c11/95	Nazzareno Abela, Zejtun Cancu Supreme
LCY 965	Volvo B10M-62	Plaxton Première 350	C53F	11/93	by 11/99	Joseph Spiteri, Ghaxaq Garden of Eden
ACY 966	Scania K93CRB	Ghabbour	C55F	11/95	c11/95	Anthony Zahra, San Gwann Zarb Coaches
LCY 967	Scania K93CRB	Ghabbour	C55F	11/95	c11/95	Angelo Zammit, Mellieha Cancu Supreme
LCY 968	Scania K93CRB	Ghabbour	C55F	c11/95	c11/95	Nazzareno Abela, Zejtun Cancu Supreme
KCY 969	Scania K93CRB	Ghabbour	C55F	c11/95	c11/95	Paul Muscat, Paola KopTaCo
ACY 970	Volvo B10M-62	Irizar Century 12.35	C53F	c11/95	c11/95	Dalli Bros., Gzira Welcome Garage
ACY 971	Volvo B10M-62	Irizar Century 12.35	C53F	c11/95	c11/95	George Caruana, Tarxien Cancu Supreme
JCY 972	Volvo B10M-62	Irizar Century 12.35	C53F	1/96	1/96	Joseph Calleja, St Julians individual owner
KCY 973	Bedford YMT	Duple Dominant II	C53F	7/80	10/86	Joseph Spiteri, Ghaxaq Garden of Eden
ACY 974	Volvo B10M-62	Irizar Century 12.35	C53F	c11/95	c11/95	Leone Grech, Mosta Paramount Garage

LCY 951, Y-2502, H930DRJ — Shearings, 930
LCY 952, Y-2503, H921DRJ — Shearings, 921
KCY 953, Y-2504, H923DRJ — Shearings, 923
JCY 954, Y-2505, H115NFX, XEL941, H650KLJ, G990FFX — Excelsior, Bournemouth
JCY 955, Y-2506, J433HDS — Redwing, Camberwell, London SE5
CCY 956, Y-2507, H617UWR — Berkeley, Paulton
JCY 957, Y-2508, H624UWR — Wallace Arnold, Leeds
ACY 958, Y-2509, K264FUV — Redwing, Camberwell, London SE5
BCY 959, Y-2510 — New
ACY 960, Y-2511 — New
LCY 961, Y-2512 — New
KCY 962,
BCY 963, Y-2514 — New
ACY 964, Y-2515 — New
LCY 965, L254UCV — Western National, 2254
ACY 966, Y-2517 — First Coaches, Malta
LCY 967, Y-2518 — New
LCY 968, Y-2519 — New
KCY 969, Y-2520 — New
ACY 970, Y-2522 — New
ACY 971, Y-2523 — New
JCY 972, Y- 2524 — New
KCY 973, Y-0909, MMJ541V — Tourmaster, Dunstable: North Star, Stevenage
ACY 974, Y-2526 — New

Paul Vella of Mosta remains as an independent coach operator, not having joined the KopTaCo fleet. KCY 969, his immaculate Volvo B10M with Irizar Century bodywork, was photographed at Cirkewwa, whilst it awaited its passengers to return from a day trip to Gozo in September 2009.

LCY 975	Volvo B10M-62	Irizar Century 12.35	C53F	c11/95	c11/95	Victor Spiteri, Luqa Century Coaches
JCY 976	Volvo B10M-62	Irizar Century 12.35	C53F	1/96	1/96	Dalli Bros. Ltd, Gzira Welcome Garage
LCY 977	Volvo B10M-62	Irizar Century 12.35	C53F	c11/95	c11/95	Joseph Zammit, Mosta Zarb Coaches
LCY 978	Volvo B10M-62	Irizar Century 12.35	C53F	c11/95	c11/95	Charles Zammit, Mosta Paramount Garage
KCY 979	Volvo B10M-62	Irizar Century 12.35	C53F	c11/95	c11/95	Paul Vella, Mosta Silver Star
ACY 980	Volvo B10M-62	Irizar Century 12.35	C53F	c11/95	c11/95	Nazzareno Abela, Zejtun Cancu Supreme
ACY 981	ERF E10 Trailblazer	Unicar	C53F	c11/95	c11/95	Emanuel Zarb, Birkirkara Zarb Coaches
LCY 982	MAN 11.190 HOCL-R	Caetano Algarve 2	C53F	c11/95	c11/95	Josephine Abela, Zejtun Cancu Supreme
LCY 983	ERF E10 Trailblazer	Unicar	C53F	c11/95	c11/95	John Galea, Balzan Arthur and John's
CCY 984	MAN 11.190 HOCL-R	Berkhof Excellence 1000 Midi	C35F	12/95	12/95	Emanuel Zarb, Gzira Zarb Coaches
ACY 985	ERF E10 Trailblazer	Camo	C53F	1/96	1/96	Kevin Muscat, Gharghur KopTaCo

LCY 975, Y-2527	New
JCY 976, Y-2528	New
LCY 977, Y-2529	New
LCY 978, Y-2530	New
KCY 979, Y-2531	New
ACY 980, Y-2532	New
ACY 981, Y-2533	New
LCY 982, Y-2535, N77YKT	New
LCY 983, Y-2536	New
CCY 984, Y-2538	New
ACY 985	New

The first coach to carry one of the new xPY registrations is Cancu Supreme's Temsa Safari HD. CPY 012 is parked in the residential area of Marsaxlokk, well away from the bustling Sunday market, in September 2009.

ACY 986	Volvo B10M-62	Plaxton Première 350	C55F	1/96	1/96	Esther Grima, Mellieha KopTaCo
xCY 987	*This registration is currently not in use*					
JCY 988	Dennis Javelin	Plaxton Excalibur	C53F (arrived 1/96)	4/97	4/97	Leone Grech, Mosta Paramount Garage
JCY 989	Dennis Javelin	Plaxton Excalibur	C53F (arrived 1/96)	4/97	4/97	Leone Grech, Mosta Paramount Garage
JCY 990	Dennis Javelin	Plaxton Excalibur	C53F (arrived 12/95)	by 4/00	by 4/00	Leone Grech, Mosta Paramount Garage
CCY 991	*This vehicle was destroyed by fire in 1/09*					Anthony Grima, Gzira KopTaCo
CCY 992	MAN 11.190 HOCL-R	Berkhof Excellence 1000 Midi	C35F	by 5/96	by 5/96	Godwin Farrugia, Mosta KopTaCo
LCY 001	Mercedes Benz OHL 1684	Irizar Century 12.35	C53FT	10/96	10/96	Leone Grech, Mosta Paramount Garage
CCY 002	MAN 11.190 HOCL-R	Berkhof Excellence 1000 Midi	C35F	by 10/96	by 10/96	Nazarene Abela, Zejtun Silver Star
LCY 002	Bedford WLB (?)	Brincat (1991)	B16R (Fomm Ir Rih)	1930	5/91	Nazzareno Abela, Zejtun Cancu Supreme
(LCY 003)	Chevrolet n/c	Brincat (1991)	B16R (Mensija) (This vehicle is stored)	1929	3/92	Nazzareno Abela, Zejtun Cancu Supreme
LCY 003	Austin CXB	Plaxton	C29F	4/50	3/07	Nazzareno Abela, Zejtun Cancu Supreme
LCY 004	Morris n/c	Brincat (1991)	B17R (Melita)	1929	10/91	Nazzareno Abela, Zejtun Cancu Supreme
CPY 012	Temsa Safari HD	Temsa	C53F	6/09	6/09	Nazzareno Abela, Zejtun Cancu Supreme
IPY 013	Temsa Safari HD	Temsa	C53F	5/09	11/09	Nazzareno Abela, Zejtun Cancu Supreme

ACY 986	New
xCY 987	
JCY 988	New
JCY 989	New
JCY 990	New
CCY 991, Y-2525	New
CCY 992	New
LCY 001	New
CCY 002	New
LCY 002, Y-0926	Origianlly, possibly, Bedford chassis
LCY 003, Y-0928, 2655, 1569	Originally, Birkirkara bus 2882
LCY 003, FCO314	Rambler, Hastings; Bruce and Roberts, Lewdown
LCY 004, Y-0927, 3295	Originally Morris Commercial chassis
CPY 012	New
IPY 013	New

COACH FIRMS ON MALTA

Garden of Eden was founded by Joseph Spiteri on leap-year day, 29th February 1932. The first vehicle operated was a Chevrolet coach with fourteen seats, numbered 2315, which was used on the route to Cospicua. This vehicle is fondly remembered by the firm and features in much of their publicity. A Diamond T route-bus with 29 seats was purchased next; number 2315 was transferred to it, as is the practice. Sometime later a Reo with a Schembri body joined the fleet. It was numbered 973 and was used as a private bus for hire. A Bedford OB, with the registration J-4792, is still stored on the company's premises. Joseph's son, Angelo, succeeded his father as director of the firm on 7th July 1968. In his free time Angelo set about building a bus for the fleet by himself. Then he converted, in-house, their Dodge route bus with Casha and then Tonna bodywork, number 2575, to a private coach. The company introduced the first minibus to Malta, which was a Commer 22-seater with a body built in-house again, and numbered 47185. When in service it carried passengers free of charge, as its dimensions contravened the law governing such vehicles. Garden of Eden continued to update the coach fleet, purchasing a number of coaches from Reliance of Newbury. Jason Camilleri, Angelo's son-in-law, recalls how he drove what became coach 863 all the way from Newbury to Reggio in southern Italy. In 1995 the company took advantage of the government's 'offer' which enabled coach owners to purchase one new bus for each bus permit owned. Consequently Dennis Javelins CCY 929 and BCY 930 were bought from England. Twelve months later two brand-new additions to

Only a few of the Garden of Eden fleet of coaches retain their distinctive artwork at the rear. JCY 899, a Bedford YMT with Duple Dominant II bodywork parked at Ghadira Bay, bears a fine view of Mosta Dome.

Paramount Garage owns LCY 860, a Dennis Javelin GX with Plaxton Excalibur bodywork, which retains its Plaxton demonstrator livery and wording. It was captured on film in the suburbs of Mosta in March 2009.

the fleet arrived in the form of an ERF Trailblazer with Unicar body, LCY 948, and a Ghabbour-bodied Scania, LCY 965. The present-day coach fleet includes a Volvo Première and Leyland Tiger, both of which have replaced their fire-damaged predecessors. Individual artwork adorns the rear window of the each coaches, with colourful paintings of some of Malta's most attractive buildings, though some have been removed during repaints. Since Angelo's death, his son Joseph has been Managing Director.

Paramount Garage was founded by Joseph Grech. Originally the company was known as 'The Assumption', a name which appeared on each of the vehicles. To begin with, during the 1930s, Joseph operated a bus, registration 3217, on the Cospicua to Valletta route. Then, in 1944, he purchased a permit for the Birkirkara route and converted a military truck into a bus. But then one of Joseph's relatives, on his return from New York, saw this newly painted bus (which was already sign-written with The Assumption!), told Joseph that "This bus is super. You should call it Paramount, because it is the best." The decision was made to adopt this name for the company. The sign-writer charged £2 to undertake the repainting, Joseph recalled. Joseph was then requested to operate a school service for pupils who lived at Mgarr. Gradually the firm increased these schools services with Paramount maintaining many for British forces stationed on the island without tenders being invited, as the service provided was known for its reliability and punctuality. Each of the school services was allocated a route number, and this practice proved to be the forerunner of the system eventually adopted for the route buses on the island, when the colour-coding for the routes was abandoned and all the buses were

Malta and Gozo Buses

painted green. On his retirement, Joseph handed over the running of the firm to one of his two sons. Leo, who now oversees the operation of eight route buses, four of which are the quartet of Dennis Dart buses on the island, fourteen coaches, minibuses and chauffeur-driven cars.

Cancu Supreme Travel is owned by the Abela family and is based in Zejtun. 'Cancu' is the family nickname. The firm's original vehicle, a Ford V8 with a Barbara body, carried the registration number 1754. In 1979, Nazzareno Abela, the current owner, bought his very own first bus. From this humble beginning, as a simple owner-driver, he managed to buy three more buses within the next few years. When he took over the family business, he entered the coaching field and registered the company name in 1984. CST imported and used to own the only Optare Excel route bus on the island which was the first low-floor bus in Malta. At the end of July 2001 CST received the first two of the new generation of vehicles, a Chinese-built King Long low-floor bus and a King Long coach. Over the years CST has evolved into the largest PSV operation on the Maltese islands. The fleet now comprises some thirty-five luxury coaches, eight route buses, four vintage buses, several mini-buses, mini-vans and chauffeur-driven private hire cars. Around twenty open-top buses are also owned. CST is also associated with a 'water taxi' service as well as cruise and harbour tour operations. **Oxley Coaches** has been absorbed into the CST fleet since the first edition of this *Bus Handbook*.

Zarb Coaches is now based in Iklin and was founded in 1947. The firm originally operated coaches with Maltese bodywork on Leyland or AEC chassis imported from Britain. When government regulations were eased, Zarb were quick to realise the opportunity of importing vehicles from Britain, and vehicle types such as AEC coaches with Plaxton bodywork appeared on the island in Zarb colours. The company is now primarily a coach firm, but a few route buses used to be operated too, including an 11-metre coach rebuilt as a 10-metre route bus. The full-size coaches from **First Coaches** were absorbed into the fleet in 2007. The original livery of orange with a single blue and white stripe has been replaced in recent years by the current scheme of white with blue flashes and an orange Zarb Coaches name. Some vehicles carry the "Sharing Holidays" logo.

KopTaCo was formed in 1997 as a co-operative of some of Malta's individual coach owners or smaller coach firms to enable them to consolidate resources. The constitution of the co-operative stipulates that work and revenue is shared equally between members. The company is dependent on tourist work and school contracts. The company livery is white with standard purple logos, unless a specific contract demands an individual livery, such as the current "Meeting Point" contract.

Companies absorbed into KopTaCo in recent years include **Morin Coaches** and **Ventura**.

Welcome Garage (Dalli Bros) was founded after the Second World War by Carmelo Dalli, initially starting with just three vehicles. Charlie, Freddie and Antoine, his sons, took over the business following their father's death in 1994. Their vehicles were replaced between 1986 and 1997 with three different types of Plaxton-bodied imports. The fleet increased to the current five vehicles when two Irizar-bodied coaches were purchased. Three red minibuses are also operated.

Arthur and John's was founded about 35 years ago. Five coaches are owned and driven by the four members of the Galea family of Balzan. The fleet livery has always been black and red.

Other individual coach firms on Malta are **Bonu Coaches, Nimrod, Peppin Garage, Schembri Coaches, Silver Star** and **Sultana**.

THE HISTORY OF BUSES ON GOZO

The early history of public transport on the island of Gozo was restricted to the use of the traditional horse-drawn carriage, called a *Karrozzin*. The first Gozo bus was a Ford imported from the USA by Loreto Xuereb in January 1925 and registered as 140. This 'matchbox' type bus conversion lasted until 1934 and was replaced by a Dodge fitted with an Aquilina body in 1935. An application to transfer it to the private-hire fleet was refused so it then was transferred to Malta where it operated on the Gudja to Ghaxaq route.

As the islands were British colonies, it was logical that English-sourced vehicles eventually appeared on Gozo, with some originating from the Army after 1945. These military vehicles were converted into buses. Other buses of American origin were brought to Gozo, too. During the Second World War, services continued to function as circumstances allowed, but they were reduced or restricted according to the availability of fuel, and some routes were cancelled. A well-known story, still often told, recalls how driver Joseph Caruana had left his bus in Sabina Square in Victoria, when trying to run away from a bomb dropped near his bus. He was wounded, but the bus remained intact.

In 1945 Joseph, nicknamed "Kelies", because he had owned a horse-drawn carriage before the war, bought a Federal, numbered 1250. The bus had 24 seats, and a rear entrance. When on duty, Joseph wore a cap and if he was caught without it, he was heavily fined. Even travelling without a conductor brought fines, and the vehicle could also be "garaged" for up to a maximum of eight days. Joseph's bus was one of fourteen on the island. During the early years after World War Two, these buses were each painted a different colour to indicate the route and village served. The bus based at Victoria served the harbour at Mgarr, Marsalforn on the north coast, Ta' Pinu and Xlendi. It boasted a grey, white and red livery. The Nadur bus was green, Xaghra red, with the owners choosing the colour schemes for themselves. The bus allocated to one route was prohibited from operating on any other.

The fourteen buses were reduced to twelve after the Government intervened in the early 1950s, stating that there two too many buses on the island. Lots were cast to decide whose vehicles would be taken off the road and sold to Maltese owners. Joseph had to sell one of his buses, as did the Nadur driver.

During the 1950s there were no foreign tourists coming to Gozo, merely local migration between the villages. Bus travel was the principal form of transport for the Gozitans, as private cars were virtually unknown. A bus owner/driver had to endure long working-hours, beginning at 0500 and often working until 2300, but he was able to make a good living. Fares varied from route to route, according to the distance from Victoria of the village served. In 1959 the bus owners formed an Association on the island. From then onwards the routes were operated on a roster basis. The various colour schemes for the buses were discontinued and the colour scheme of the Victoria bus (grey, with a white roof, and one red band at waist level) was chosen for the island's whole fleet, a colour scheme which continues to this day. Local private work on the island was based on free enterprise.

In the early 1960s a Sunday service from Xaghra to Victoria was introduced, starting at 1600 and finishing at 2100, to cater for local residents wishing to go to the cinema. This service remained unique to Xaghra for most of the decade, but during the 1970s most of the other villages boasted similar services. By the early 1980s these services had been withdrawn, however, the passage of time having been marked by the increasing number of private cars on the island, owned especially by the younger generation.

Bus services on Gozo were not centralised on the present-day bus-station in Victoria until 1965. Until then individual routes boasted their own termini around the capital. The

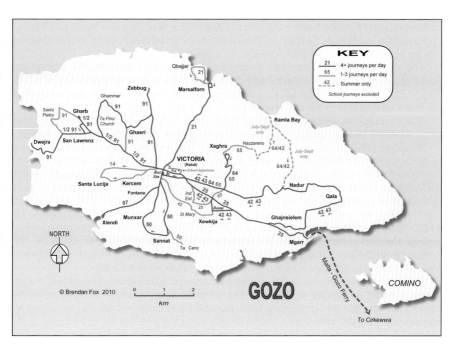

terminus for the route to Mgarr was Independence Square; to Xaghra from near the Police Station; to Nadur, Xewkija, Qala and Marsalforn from Main Gate Street; to Xlendi from St Augustin Square; to Sannat from St Francis Square and to Gharb, San Lawrenz, Ghasri and Zebbug from Sabina Square. When the new bus station in Triq Putirjal was opened, the bus drivers soon went on strike, as their vehicles would not easily start on the upward hill towards St Francis Square.

The number of buses on the island gradually increased over the years. The original Gozo buses included a Federal, an International, a Reo and a Ford ET7 forward-control bus, each with a body by a Maltese builder, such as Borg, Casha or Debono. During the 1960s the number rose to eighteen, when some Bedford chassis with Leyland engines were bought new and were bodied, just as their sister vehicles on Malta, by Aquilina. A further ten buses arrived in 1974, some of which were buses transferred from Malta. During the mid-1980s vehicles imported from Britain appeared on the island. Two Bristol LHs preceded a number of Dominants, one Ford Plaxton and a Hino which were imported – either from Britain or from Malta — after the Government relaxed the laws limiting the size of vehicles. So, buses of 10 metres in length were permitted on the island's roads. The names of the lucky drivers receiving these vehicles were drawn by lot. Five years later another three vehicles were introduced, after a computer selection of the three new owners. This brought the total of buses on the island to thirty-nine. In 1996 the Government gave every driver permission to buy one new vehicle for each one he already owned to cater for the increase of tourism on the island.

The present-day total of 78 buses and coaches still includes many of the vehicles involved in the development of the island's bus fleet outlined in the preceding paragraph. The doubling of the fleet brought a large number of coaches on Optimal chassis with bodies built on Malta by Scarnif. An array of distinctive demonstrator buses and coaches appeared too, on British, German or French chassis with bodies built by a wide choice of foreign builders, from Italy and Portugal, for instance.

The original buses on Gozo had seating for 32, 34 or 36 passengers. The present-day fleet consists principally of vehicles with 36, 45 or (for the coaches) 53 seats. Up to twelve standing passengers are allowed on route buses, with none on the coaches. Registrations of the buses on Gozo have always been based on the series current in Malta.

There are eight buses operating the island's routes each day, with one (or two) vehicles being allocated to one specific route, or a series of interworked routes throughout the day.

Services to the villages run at a loss nowadays. But there is still a reasonable provision to all the principal villages on the island, with special journeys for school-children during the term-time to and from primary schools and the secondary school in Victoria. During the summer months Marsalforn and Xlendi, in particular, have services every 45 or 60 minutes throughout the day. The timings to Marsalforn in summer 2009, for instance, totalled 24 a day, with extensions along the promenade to Qbajjar every two hours. The small villages of Santa Lucija and Kercem, on the other hand, have three early morning timings and then nothing else. Unique amongst the Gozo routes nowadays, this route does not start from the Bus Station, but from Sabina Square. Services from Victoria to the ferry at Mgarr leave thirty minutes before a ferry departure throughout the day.

With Gozo being so small and Victoria being situated in the centre of the island, journey times on most routes are very short. The timetables at the main bus-station advise passengers that the return journey from the outer terminus will usually be about ten minutes after the departure time from Victoria. Exceptions to this are to be found on the almost hour-long route 91 to Gharb, Dwejra and Zebbug, or the summer extension of route 42 serving Nadur and Ramla Bay.

Joseph Caruana's son, Emanuel, is the longest serving owner-driver on the island, having worked for forty-two years. When he was ten years old, he started as a conductor for his father and was paid the equivalent of £1.50 a week. On his seventeenth birthday in 1967 he received his bus driver's licence. He can laugh now about these early days, but when he first became a driver, regular passengers queried his ability and qualification to do so at such a tender age. To begin with he drove his father's bus, but in 1982, when he was 32 years old, he acquired his own first bus which he had to buy in the name of his fiancée, Mary Mifsud. Legislation at that time forbade bus-owners purchasing a second vehicle in their name and Emanuel was still co-owner of his father's bus. Emanuel, his brother Michelangelo and nephew, John, continue as drivers on the island. Another brother, Anthony, retired in 2002, aged 61 after driving for thirty-four years and sold his buses to Paul Borg. In recent years other owner-drivers have retired; Philip Bonnici of Sannat sold his vehicles to Mariano Farrugia and John Portelli to Emanuel Custo.

Owners are proud of their vehicles which are kept in immaculate condition. Each of the 78 buses is unique, not necessarily because of its vehicle type, body or chassis, but because of the application of the red band in its colour scheme, or its individual wording on windscreen or bodywork. Emanuel Cutajar's Bedford (FBY 032) proclaims SPEEDWAY GARAGE on its front windscreen; similarly, Joe Xerri's Bedford (FBY 061) proudly and incongruously shows GEORGE EDWARDS AND SON on the front windscreen and Coachmen since 1923 at the rear, thus still indicating the former owner of the coach near Wrexham, north Wales; Optimal/Scarnif FBY 078 has SECRETS to reveal, whilst a similar vehicle FBY 075 shows LEPEIRKS Coaches, which is merely the nickname of the owner, Skriepel, spelt backwards. Amongst the older vehicles the Reo (FBY 042) shows its nickname DOLLY, and Bedford SB (FBY 038) shows CLOUDS. Many drivers are employed, some part-time, by the owners nowadays. As on Malta, many of the vehicles are garaged overnight at the homes of their owners in villages throughout the island. Consequently, the daily timetable of the

Rambo, FBY 043, is one of the longest-serving vehicles on the island. Its Dodge chassis dates from 1942 and its Brincat bodywork was completed in June 1968. It is seen entering Sannat from the Munxar direction on an early morning school journey in October 2007. These morning school journeys work in the opposite direction to normal daily public timings.

final departure from Victoria of some of the routes bears the unusual proviso "This service does not return to Victoria". Advertisements are still prohibited on Gozo's buses and whilst owners are allowed to show their company name and contact phone numbers, very few of them do so, as word of mouth is far more effective on the island, where everyone knows one another.

Bus routes on Gozo remain very static, with the only significant alteration since the first edition of the *Malta Bus Handbook* being the unheralded withdrawal in the mid-2000s of the afternoon departures from the Industrial Estate situated on the eastern outskirts of Victoria. Timetables of the individual routes do vary according to season and an explanation of these seasonal changes accompanies each route description in a later chapter.

A recent innovation, which, sad to say, lasted only for the duration of its Pilot Project was the "Discover Gozo by Bus" service. This operated as a "Hop-on Hop-off Service or Round Trip" (as the advertisement proclaimed) between 16th July and 5th August 2008. Organised by the Mediterranean Standard for Sustainable Tourism, this project provided a round trip of the island, giving passengers the opportunity to spend a while in different villages on the island, thus waiting for the next bus to arrive. The service operated at two-hourly intervals from Sabina Square in Victoria between 0930 and 1730.

The timetable for the first timing of the day was as follows: Victoria Sabina Square 0930; Mgarr (up the hill from the ferry) 0945; Nadur 1005; Ramla Bay 1020; Ggantija Temples (Xaghra) 1040; Marsalforn 1100; Ta' Pinu 1120; Dwejra 1140; Xlendi 1205; Victoria 1220. Subsequent departures followed this pattern at two-hourly intervals. The service was not introduced for the 2009 season, despite Gozitan bus drivers and local officials meeting to discuss the success and future of the project in the weeks immediately following the 2008 trial period. Another short-lived bus service with Gozo connections was the Gozo-Airport Shuttle service.

VEHICLES ON GOZO

Registration	Chassis	Bodywork	Seating	New	to Gozo	Owner
FBY 001	Bedford YNV Venturer	Caetano Algarve N-NDH	C53F	3/88	5/97	Carmel Cauchi, Xewkija
FBY 002	MAN 11.192	Dorteller	B45F	1/93	6/97	Jason Farrugia, Victoria
FBY 003	Iveco 380	Cacciamali	B45D	by 12/96	6/97	Teddy Farrugia, Victoria
FBY 004	Dennis Javelin 11SDL1905	Plaxton Paramount 3200 III	C53F	7/88	6/97	Peter Paul Debono, Nadur
FBY 005	Mercedes Benz 1310/50	Castrosua	B45F	by 1/91	6/97	George Farrugia, Victoria
FBY 006	ERF E6.18BC2	Marshall Islander	B45F	4/95	6/97	Josmar Farrugia, Victoria
FBY 007	Dodge f/c	Aquilina	B45F	3/70	10/71	John Attard, Xaghra
(This vehicle passed to its present Gozo owner in 4/89, yet received its FBY 007 registration only in 6/97, and has been in store since 2002)						
FBY 008	Dennis Javelin 11SDA1906	Plaxton Paramount 3200 III	C53F	8/89	8/97	Emanuel Cutajar, Victoria
FBY 009	Sanos S315-21	Sanos Carisma	C53F	5/90	6/97	Joseph Portelli, Gharb
FBY 010	Dennis Javelin 11SDL1905	Plaxton Paramount 3200 III	C53F	9/87	8/97	George Tabone, Fontana
FBY 011	Bedford YNT	Plaxton Paramount 3200 III	C53F	10/88	6/97	Emanuel Mintoff, Ghasri
FBY 012	Optimal	Scarnif	C53F	9/97	9/97	Andrew Custo, Gharb
FBY 013	Dennis Javelin 11SDL1905	Plaxton Paramount 3200 III	C53F	10/88	8/97	Teddy Cassar, Victoria
FBY 014	Dennis Javelin 12SDL	Plaxton Paramount 3200 III	C53F	3/88	by 3/98	Joseph Custo, Gharb

FBY 001, 6541FN, E348TPW — Cobham, Great Yarmouth
FBY 002 — Maltese demonstrator
FBY 003 — Maltese demonstrator
FBY 004, E538PRU — Lewis, Llanrhystyd: Moon, Warnham
FBY 005 — Maltese demonstrator
FBY 006 — Maltese demonstrator
FBY 007, Y-0892, Y-1592, A-7245, 7245 — Unscheduled bus, Malta
FBY 008, G778APK — Bicknell, Godalming
FBY 009, G895VNA — R. & N. Lyles, Batley: Shearings, 895
FBY 010, E839EUT — Bicknell, Godalming: Briscoe, Whitley
FBY 011, HIL2385, D390BNR — Hearn, Harrow Weald: Owen, Yateley
FBY 012 — New
FBY 013, F900RFH — Stevens, Bristol: Pulham, Bourton-on-the-Water
FBY 014, E514WAH, WOA521, E511JWP — Sanders, Holt: Whittle, Highley

School is out! The mass exodus of pupils at 1430 every school day is not to be missed. Every vehicle in service on that day is rostered to provide a journey home for pupils of the senior college in Victoria. In March 2009, FBY 014, a Dennis Javelin with Plaxton Paramount 3200 III bodywork, waits for its pupils to arrive.

FBY 006 has a Marshall Islander body on an ERF lorry chassis and is one of the demonstrator vehicles which were originally delivered to Malta for evaluation purposes in the mid-1990s. It is seen on lay-over in Victoria after working route 65, the extension of the normal route 64 to Tan Nazzarenu on the outskirts of Xaghra. Now owned by the Farrugia family in Victoria, it still betrays half of its former identity of 'Bonnici Coaches' in its destination board.

FBY 015	Bedford SB1	Zammit	B36F	1954	7/64	Michaelangelo Caruana, Xaghra
FBY 016	Bedford YRQ	Plaxton Panorama Elite III	C45F	9/74	3/90	George Farrugia, Victoria
FBY 017	Bedford YNT	Plaxton Paramount 3200 I	C53F	3/85	6/93	Emanuel Mintoff, Ghasri
FBY 018	Bedford YLQ	Duple Dominant II	C45F	3/78	7/88	Joseph Custo, Gharb
FBY 019	Bedford SBG	Aquilina (1972)	B36F	3/57	5/72	Andrew Custo, Gharb
FBY 020	AEC Reliance 6U3ZR	Duple Dominant II	C53F	5/79	by 10/91	Mariano Farrugia, Xewkija
FBY 021	Ford ET7 f/c	Borg (rebuilt 1978)	B40F	4/53	4/53	Emanuel Cutajar, Victoria
FBY 022	Leyland Comet	Aquilina	B36F	5/64	8/68	Mariano Farugia, Xewkija
FBY 023	Bedford YRQ	Duple Dominant I	C45F	3/75	12/87	Raymond Pace, Kercem
FBY 024	Leyland Leopard PSU3/1R	Plaxton Supreme II (1977)	C53F	5/62	3/86	Emanuel Tabone, Munxar
FBY 025	Bedford SB8	Aquilina	B36F	2/62	2/62	Saviour Buttigieg, Qala
FBY 026	Bedford SB8	Aquilina	B36F	5/63	5/63	John Attard, Xaghra
FBY 027	International Harvester K f/c	Casha (rebuilt 1966)	B36F	1942	2/50	George Tabone, Fontana
FBY 028	Hino BT51	Debono	B36F	by 9/74	2/77	Teddy Cassar, Victoria
FBY 029	Bedford YLQ	Duple Dominant II	C45F	5/79	7/88	Ennio Farrugio, Victoria
FBY 030	Bedford YRQ	Duple Dominant I	C45F	9/75	by 4/86	Ludovico Azzopardi, Zebbug
FBY 031	Bedford YMT	Duple Dominant I	C53F	6/76	by 3/86	Joseph Grima, Qala

FBY 015, Y-0827, Y-1528, A-3095, 3095	To Gozo on 15/2/74: route bus, Malta on an RN chassis
FBY 016, Y-0821, Y-0466, UUP2N	Route bus, Malta: Armstrong, Ebchester
FBY 017, Y-0845, B493BJO, 499BHU, B730YNM	Majestic, Shareshill: Beckett, Little Horwood
FBY 018, Y-0841, AGG934S	Chapman, Airdrie: Strathmore Coaches, Dundee
FBY 019, Y-0818, Y-1519, A-2483, 2483	British Army, 44 BS 45 with Mulliner B39F bodywork
FBY 020, Y-0820, BHL472T	Ash, Woodburn Moor: Kirkby, Sheffield
FBY 021, Y-0824, Y-1525, A-3062, 3062	New
FBY 022, Y-0830. Y-1531, A-3184, 3184	New
FBY 023, Y-0839, HSG724N	Bland, Stamford: Silver Fox, Edinburgh
FBY 024, Y-0809, Y-0897, OBN534R	To Gozo by 4/91: Unscheduled bus, Malta (Y-0897)
FBY 025, Y-0832, Y-1535, C-1503, 31503	New
FBY 026, Y-0825, Y-1526, A-3065, 3065	New (This vehicle has been stored since 2002)
FBY 027, Y-0810, Y-1511, A-1126, 1126	Route bus, Malta: RAF bowser, F6103.
	Purchased by dealer 10/49
FBY 028, Y-0840, Y-0888, Y-1588, A-2362, 2362	To Gozo in 5/88: Unscheduled bus, Malta (Y-0888)
FBY 029, Y-0842, YRG702V	Glyn Evans, Manmoel
FBY 030, Y-0813, KPC213P	Reliance, Newbury, 157: Gale, Haslemere
FBY 031, Y-0831, LTF224P	Reliance, Newbury, 147

FBY 044 is a recently imported coach and is one of a few vehicles on the islands which boast an Irish heritage. This Scania K124 with Irizar bodywork saw service both in Ulster and Eire before arriving on Gozo and it is seen outside George Farrugia's premises in Victoria in March 2009.

FBY 032	Bedford YNT	Plaxton Paramount 3200 II	C53F	3/86	7/93	Emanuel Cutajar, Victoria
FBY 033	Bedford YRQ	Duple Dominant I	C45F	2/76	5/88	Joe Xerri, Xaghra
FBY 034	Bedford YMT	Duple Dominant II	C53F	5/77	by 6/00	Mary Caruana, Victoria
FBY 035	Bristol LH6L	Eastern Coach Works	B45F	12/74	by 9/85	Godfrey Borg, Victoria
FBY 036	Ford R1014	Plaxton Supreme IV Express	C45F	6/81	5/88	Joseph Custo, Gharb
FBY 037	Leyland Royal Tiger PSU1/15	Schembri (1973)	B40C	5/53	6/73	Peter Paul Debono, Nadur
FBY 038	Bedford SBG	Debono (1968)	B36F	by 12/56	12/68	George Farrugia, Victoria
FBY 039	Bedford YNT	Plaxton Paramount 3200	C53F	3/84	7/93	Teddy Farrugia, Victoria
FBY 040	Bedford SBO	Aquilina	B36F	4/56	11/56	George Farrugia, Victoria
FBY 041	Bedford YLQ	Duple Dominant II	C45F	2/79	7/88	Jason Farrugia, Victoria
FBY 042	Reo f/c	Zammit (1962)	B36F	by 12/33	by 12/33	George Farrugia, Victoria
FBY 043	Dodge T110L f/c	Aquilina (7/54); Brincat (6/68)	B36F	1942	7/54	Jason Farrugia, Victoria
FBY 044	Scania K124	Irizar	C53F	4/00	9/08	Ennio Farrugia, Victoria
FBY 045	Bedford YMT	Duple Dominant I	C53F	4/79	by 8/86	Carmel Cauchi, Victoria
FBY 046	Bedford YNT	Plaxton Paramount 3200 III	C53F	8/87	9/97	Emanuel Caruana, Victoria

FBY 032, Y-0846, D354CFR, BIB7667, C212DEC — Bibby, Ingleton
FBY 033, Y-0838, MRB686P — Clarke, Newthorpe
FBY 034, LCY 860, Y-0860, SBC 57R — Unscheduled bus, Malta (LCY 860):
County Travel, Leicester, 57
FBY 035, Y-0811, Y-1512, GLJ482N — Hants and Dorset, 3550
FBY 036, Y-0837, MJX791W — Stopps, Uxbridge: Terminus, London W1
FBY 037, Y-0815, Y-1516, A-2070, 2070, FCK402 — Wimpey, London W6: Ribble, 902
FBY 038, Y-0817, Y-1518, A-2377, 2377 — RAF, 32 AC 67 with Mulliner B31FA bodywork
FBY 039, Y-0844, A158DWP, JPY505, A468MRW,
9258VC, A838PPP — Whittle, Kidderminster: Armchair, Brentford
FBY 040, Y-0828, Y-1529, A-3126, 3126 — New
FBY 041, Y-0843, JTU227T — Bostock, Congleton, 26
FBY 042, Y-0816, Y-1517, A-2296, 2296 — To Gozo in 7/73: route bus, Malta. Chassis new n/c in 1933
Rebuilt f/c in 1962
FBY 043, Y-0822, Y-1523, A-2844, 2844 — To Gozo on 9/5/76: route bus, Malta. Chassis new 1942
as military lorry n/c and rebuilt to f/c in 1968
FBY 044, INZ5104, 00-KK-2664 — McFadden, Cookstown, Northern Ireland:
Kavanagh of Urlingford, Co. Kilkenny, Eire
FBY 045, BCY916, Y-0916, FKX275T — Unscheduled bus, Malta (BCY 916): Cavalier, Hounslow
FBY 046, SJI7046, E848KCF, OVK902, E829EUT — Venture, Harrow: Robin Hood, Rudyard

FBY 051 is another of the venerable Maltese buses. This Bedford SB with a body built by Aquilina now boasts well over forty-five years in service on the island. It is seen on the outskirts of Marsalforn on route 21.

FBY 047	Bedford YNT	Plaxton Paramount 3200 III	C53F	6/87	9/97	Raymond Pace, Kercem
FBY 048	Bristol LH6L	Plaxton Supreme II	C45F	1/78	4/86	Mariano Farrugia, Xewkija
FBY 049	Bedford YNT	Plaxton Supreme V	C53F	10/81	12/93	Joseph Portelli, Gharb
FBY 050	Bedford YLQ	Duple Dominant Express II	C45F	4/79	by 7/88	Teddy Farrugia, Victoria
FBY 051	Bedford SB8	Aquilina	B36F	2/62	2/62	Carmel Galea, Xaghra
FBY 052	Hino BT51	Debono	BC45F	by 9/74	12/76	John Caruana, Victoria
FBY 053	Bedford SB8	Aquilina	B36F	4/62	4/62	Carmel Cini, Xewkija
FBY 054	Bedford YMT	Plaxton Supreme III	C53F	6/76	3/86	Paul Borg, Victoria
FBY 055	Bedford YNT	Plaxton Paramount 3200 III	C53F	5/87	10/97	Godfrey Borg, Victoria
FBY 056	Dennis Javelin	Plaxton Paramount 3200 III	C53F	1/89	9/97	Emanuel Tabone, Munxar
FBY 057	Maltese f/c (Zammit)	Barbara	B36F	8/69	8/69	Jason Farrugia, Victoria
FBY 058	Dennis Javelin	Duple 320 Express	C53F	4/88	by 11/97	Jason Farrugia, Victoria
FBY 059	Bedford YNV Venturer	Duple 320	C53F	3/88	9/97	Teddy Farrugia, Victoria
FBY 060	Bedford YNT	Plaxton Paramount 3200 III	C53F	8/88	10/97	Carmel Galea, Xaghra
FBY 061	Bedford YNT	Plaxton Paramount 3200 III	C53F	8/87	9/97	Joe Xerri, Xaghra
FBY 062	Dennis Javelin	Plaxton Paramount 3200 III	C53F	10/91	by 4/98	Ludovico Azzopardi, Zebbug

FBY 047, HIL 3478, D865ATF — Hill, Congleton: Station Car Hire, Binfield
FBY 048, Y-0829, WJN22S — Stowell, Fearnhead: Boon, Boreham
FBY 049, Y-0814, Y-0889, VPR864X — Gozo in 1993: unscheduled bus, Malta (Y-0889): Reliance of Newbury, 176: Tedd, Thruxton. Imported 2/86
FBY 050, Y-0836, OPT471T — Tindall, Low Fell: Langley Park Motor Company
FBY 051, Y-0835, 31504 — New
FBY 052, Y-0823, Y-1524, A-3017, 3017 — New (in service only by 9/79)
FBY 053, Y-0833, Y-1534, C-1505, 31505 — New
FBY 054, Y-0812, PGT527P — County, Brentwood: Margo, Thornton Heath
FBY 055, D614GDU, 4828VC, D677FWK — Roy Brown, Builth Wells: Vanguard, Bulkington
FBY 056, F386MUT — Brodyr James, Llangeitho: Green, Kirkintilloch
FBY 057, Y-0819, Y-1520, A-2522, 2522 — Gozo in 2/74: Malta route bus on new chassis imported from GB
FBY 058, GIL8488, E38SBO — Stone, Aldershot: Bebb, Llantwit Fardre
FBY 059, E252PEL — Richards Bros., Cardigan: Seaview, Parkstone
FBY 060, KSU363, F327YTG — Brylaine, Boston: East Glamorgan, Nelson
FBY 061, E370ECJ — George Edwards & Son, Bwlchgwyn
FBY 062, J733KBC — Warrington, Ilam

One of the Maltese-built coaches on Gozo is FBY 079, an Optimal with Scarnif bodywork. It is parked outside the premises of George Farrugia in Victoria. The variation in style of the red and white striping in the Gozo livery is used to good effect.

FBY 063	Sanos S315-21	Sanos Carisma	C53F	5/90	6/97	Carmel Cini, Xewkija
FBY 064	Dennis Javelin	Duple Metsec	B45F	1993	by 5/98	George Farrugia, Victoria
FBY 065	Bedford YNV Venturer	Caetano Algarve N-NDH	C53F	3/87	1/98	Michelangelo Caruana, Xaghra
FBY 066	TAZ D3200	TAZ Dubrava	C53F	5/90	by 4/98	Paul Borg, Victoria
FBY 067	Bedford YNT	Plaxton Paramount 3200 III	C53F	5/87	3/98	George Farrugia, Victoria
FBY 068	Leyland Tiger TRCTL11/3RZ	Plaxton Paramount 3200 III	C53F	5/87	1/98	Ennio Farrugia, Victoria
FBY 069	*This registration has never been allocated.*					
FBY 070	Optimal	Scarnif	C53F	5/98	5/98	Andrew Custo, Gharb
FBY 071	Optimal	Scarnif	C53F	6/98	6/98	Joseph Grima, Qala
FBY 072	Dennis Javelin	Duple 320	C53F	2/89	by 11/98	Joyce Farrugia, Victoria
FBY 073	Dennis Javelin	Duple 320 Express	C53F	4/88	by 2/99	Teddy Farrugia, Victoria
FBY 074	Optimal	Scarnif	C53F	8/98	8/98	Mariano Farrugia, Xewkija
FBY 075	Optimal	Scarnif	C53F	8/98	8/98	Rita Farrugia, Xewkija
FBY 076	Bedford YNV Venturer	Duple 320	C53F	4/87	by 2/99	John Caruana, Victoria
FBY 077	Optimal	Scarnif	C53F	11/98	11/98	Emanuel Custo, Gharb
FBY 078	Optimal	Scarnif	C53F	11/98	11/98	Saviour Buttigieg, Qala
FBY 079	Optimal	Scarnif	C53F	11/98	11/98	Mariano Farrugia, Xewkija
IPY 012	King Long XMQ6900	King Long	C30F	9/09	9/09	Ennio Farrugia, Victoria

FBY 063, G894VNA	R. & N. Lyles, Batley: Smiths Shearings, 894
FBY 064	Maltese demonstrator
FBY 065, D511WNV	Ham, Flimwell: Alpha, Brighton
FBY 066, G579WUT	Smith, High Wycombe: Eurobus, Harmondsworth
FBY 067, D854XYJ	Trevan, Crawley
FBY 068, D64MWO	Ham, Flimwell: Hills, Tredegar
FBY 070	New
FBY 071	New
FBY 072, 428EJB, F906UPR	Aron, Northolt: County of Avon, Bristol
FBY 073, GIL8487, E32SBO	Quantum, Swadlincote: Bebb, Llantwit Fardre
FBY 074	New
FBY 075	New
FBY 076, DIL7916, D121EFH	Hertfordshire Travel, Park Street : Swanbrook, Cheltenham
FBY 077	New
FBY 078	New
FBY 079	New
IPY 012	New

ROUTE 1 OR 2: VICTORIA to GHARB/ST LAWRENZ
ROUTE 91: VICTORIA to DWEJRA

This trio of routes covers the villages in the west of Gozo. On leaving Victoria and negotiating the sharp left-hand bend on the outskirts of the town, the bus soon passes the remains of an old aqueduct built in the 1840s by the British to carry water from Ghar Ilma to the capital. Then Ta' Pinu Church situated along a road to the right can be seen. Gharb village is soon reached – its name means "west" – and it is the westernmost village on the island. The bus turns round in the village square, where there is a typical British telephone box outside the police station. The square is dominated by the imposing parish church, designed according to the plans of Francesco Borromini's Sant' Agnese in Agone at the Piazza Navona in Rome. This Baroque church is dedicated to Our Lady's Visitation and was built in 1670, and dedicated fifty years later. It has two belfries with one displaying "Ibni ghozz iz-zmien" (Cherish time, my son). Symbols of Faith, Hope and Charity are carved around the main door. The village is home to the island's Folklore Museum, near the church in the village square. The bus then skirts the village and heads for San Lawrenz, where the Ta' Dbiegi Crafts Village is situated. Built on the site of a former British army camp, this attraction is the place to buy local pottery, glass and lacework. San Lawrenz was the home of the novelist Nicholas Monsarrat during the 1970s and is where he wrote his novel *The Kappillan of Malta*.

Holidaymakers make full use of this route, as the bus then travels down a winding road offering wonderful sea views on the approach to Dwejra Bay. Here boats are available to take tourists across the Inland Sea, through the natural tunnel in the cliffs to the Azure Window and Fungus Rock, the natural wonders of the island. Fungus Rock gains its name from the plant, *Cynomorium coccineus*, which grows on it. The plant is native to North Africa and Fungus Rock is unique, being the only place in Europe where it grows. The Knights Templar believed that the fungus cured dysentery and ulcers. They used it to staunch wounds and to treat blood diseases and even sent it as a princely gift to European kings and queens. Fungus Rock was closely guarded by the Knights and it could be accessed only by a hoist from the nearby Qawra Tower. The Azure Window is a natural limestone archway and the Inland Sea boasts aquamarine water.

The bus climbs back up to San Lawrenz, and then, depending on the timings, may serve San Pietru and Ghasri and Ta' Pinu Church, built between 1920 and 1931. This basilica with its single campanile, 47m high, is the national shrine to the Blessed Virgin. The site has been a place of worship since the early sixteenth century. As early as 1575 the church was condemned as unsafe for worship, yet it was never razed to the ground, nor wholly repaired. The last effort to improve the building was undertaken in the late seventeenth century by the family of Filippino Gauci, who was nicknamed Pinu for short. This old church is now sited in a small chapel behind the apse in the main church.

The basilica owes its fame to Carmela Grima, who, on 23rd June 1883 heard a voice asking her to say three Hail Marys "in memory of the three days My Body rested in the sepulchre". One other local resident, Francesco Portelli said that he, too, had heard voices. During the following year, Gozo escaped the great plague which attacked Malta, and consequently miracles and narrow escapes were then attributed to the grace and favour of Our Lady of Pinu. The church was raised to the status of a basilica by Pope Pius IX in 1932. Opposite the church is a Way of the Cross up Ta' Ghammar Hill with fourteen life-size marble statues representing the Stations of the Cross.

FBY 005 is unique on the islands. It is a Mercedes Benz vehicle with Castrosua bodywork and was one of the demonstrators used on Malta in the 1990s, before entering service on Gozo. It has just arrived at Dwejra and the many passengers aboard are about to leave for a day at the Azure Window.

Ghasri is the smallest village on the island; its name is derived from the Arabic for "to press". Dominating the village is a lighthouse built on Gordan Hill in 1853 from where its beam can be seen almost fifty kilometres away.

A minor road then brings the bus to the hill up to Zebbug, a village whose name means "olives" in Maltese, which, along with Ghasri's name, has led historians to suggest that the Romans and Arabs had a thriving oil industry in the area. On a slope of one of the nearby hills there is a deposit of onyx. The village church is dedicated to Santa Marija. The local inhabitants are famed for their lace-making.

The bus turns, once again in the village square in front of the church, and returns the way it had come back down the hill, along the main roads back into Victoria. Unlike many of the other routes on the island which reach their outer destination in ten minutes or so from Victoria, this route, when serving Dwejra, can take up to an hour.

The six departures from Victoria during the summer months of 2009 were at 0830 and 0930 which show route 1 (or 2), and are short-workings to Gharb. The remaining timings are route 91 as Dwejra is served, at 1030, 1130, 1315, 1500 and 1730 which also operate on Sundays and Public Holidays.

ROUTE 14 to ST LUCIA and KERCEM

This route boasts only three morning departures throughout the year, at 0715, 0900 and 1000 Monday to Saturday. The last two timings depart from Sabina Square at the far end of Victoria, before serving the Bus Station, as the bus has previously worked route 1 to Gharb. Kercem is the village closest to Victoria. The main road to the village passes through the upper part of the fertile Lunzjata Valley, where the Knights of St John loved to hunt. A pleasant walk through the valley reveals remains of their hunting-lodges and the small Chapel of the Annunciation which was erected in 1347 and rebuilt in the seventeenth century. It was partly built into a cave and is still in use today. The outward journey by bus, however, uses a rougher minor road to reach Santa Lucija before returning along the main road to Kercem, where there is the site of Christian catacombs from the third and fourth centuries at Ghar Gerduf and possibly even earlier Roman baths. Very much more recent developments in the area have seen an Italian company start drilling for oil.

Another venerable Gozo vehicles is FBY 053, a Bedford SB8 with Aquilina bodywork. In service since April 1962 it is well known for displaying destinations - a very rare instance on the islands.

ROUTE 21 to MARSALFORN

Marsalforn lies on the north coast of Gozo, and is nowadays the busiest of the island's resorts in the summer. Until the development of Mgarr in the seventeenth century, Marsalforn was the principal harbour for Gozo. The village church is dedicated to St Paul Shipwrecked and dates from 1730. Tradition has it that St Paul set sail from here, after his three-month stay on the Maltese islands.

The bus route to Marsalforn from Victoria descends to a plain overlooked by the strangely shaped conical hill on top of which is the Statue of the Saviour. Islanders once feared that the hill was a dormant volcano (because of its shape) and so built the holy statue on its summit so that it would not dare erupt.

Marsalforn is served every day of the week throughout the year. The winter service from 1st November until 31st March provides seven timings at 0735, 0905, 1030, 1200, 1330, 1430 and 1635, with Qbajjar served on the 1030, 1200 and 1430 departures. Qbajjar is reached along the extensive tree-lined promenade from Marsalforn village centre. A short walk from Qbajjar to Xwenji along the coastline reveals a number of salt pans of Roman origin. On leaving Qbajjar, the bus now completes a circle past the new housing estates back to Marsalforn, instead of returning along the promenade, this being a comparatively recent introduction since the first edition of this *Bus Handbook*.

During April, May, June and October the service is increased to eleven journeys, at 0735, 0905, 1000 (to Qbajjar), 1100, 1200 (to Qbajjar), 1330, 1430 (to Qbajjar), 1600, 1700, 1800 and finally 1845 (which, on some timetables is shown as operating only M-S).

During the summer months from 1st July to 30th September, bus services to Marsalforn are the most intense on the island. Three early timings at 0600, 0645 (both only M-S, except Public Holidays) and 0735 are then followed by departures every three-quarters of an hour from 0830 until 2200 and a final departure at 2230 which does not return from Marsalforn until 2300. Six of the timings, at approximately two and quarter-hourly intervals starting with the 0735 departure, extend to Qbajjar. Route 21 takes about twenty minutes for the round trip to Marsalforn, with the Qbajjar extension adding eight minutes to the journey time.

On Schooldays, the departure at 1430 leaves Vajringa Primary School before serving the bus station in Victoria, then going to the Sports Complex for the end of school for the senior pupils, then on to Qbajjar, excluding Republic Street in Victoria. The return journey via Marsalforn is a normal service back to the bus station.

FBY 019 is a Bedford SBG with Aquilina bodywork. On route 21 every third departure on the frequent summer timetable extends along the promenade from Marsalforn to Qbajjar, where the bus is seen.

ROUTE 25 to MGARR

This route is operated in accordance with the ferry departures from the harbour at Mgarr. The Gozo Channel Company provides a regular service between Mgarr and Cirkewwa on the northern tip of Malta every day. Buses leave Victoria half an hour before each ferry departure. The first bus is at 0530 which, on weekdays only, serves Xewkija and Ghajnsielem as well. The last bus back from Mgarr varies according to season, but is no later than 1945, even at the height of the summer tourist season. Two buses are rostered to provide this service each day, with any of the other six buses in service likely to be called upon to provide extra capacity if necessary, especially later on in the afternoons when the day-visitors to Gozo are returning to Malta. During the summer up to 45 departures can take place each day.

The route simply covers the main road between the two termini and takes about a quarter of an hour. From Victoria, en route to Mgarr the bus passes the Industrial Estate, an old farmhouse, now renovated as the home of Gozo Heritage, an audio-visual pageant which narrates 7,000 years of the history of the island, then the outskirts of Ghajnsielem, before the steep descent to Mgarr Harbour where the new arrival hall welcomes passengers and provides excellent indoor facilities prior to the ferries' departures.

Mgarr is nowadays the only gateway to Gozo. Overlooking the harbour are Fort Chambray which was built by the Knights of St John in 1749 and the church dedicated to Our Lady of Lourdes. The harbour itself and nearby marina are a haven for small craft and fishing boats. On the busy summer days when coach tours for holiday-makers from Malta are arranged, maybe as many as thirty of the island's coach fleet can be seen here waiting for their passengers between 0930 and 1000.

Hino FBY 028 is seen at the Mgarr terminus of route 25 from Victoria. Fort Chambery can just be seen on the hill in the background.

ROUTES 42/43 to NADUR, QALA, GHAJNSIELEM, XEWKIJA and RAMLA

These routes serve the easternmost villages of the island and, in summer, another of the popular tourist areas, Ramla Bay, on the north coast. Route numbers 42 and 43 indicate whether the villages are served in anti-clockwise or clockwise order from Victoria. From October to June there is a total of thirteen departures on the routes, whilst in the summer months (as per the 2009 summer timetable) there are fifteen departures, at half-hourly intervals from 0730 until 1100 (which is the first of the day to serve Ramla), then an 1145 departure which finishes at Nadur, and so does not return to Victoria. During the afternoon, two timings at 1315 and 1500 serve Ramla, followed by early evening departures at 1700, 1800 (to Ramla Bay too) and 1900.

From Victoria, the first village to be served on the anti-clockwise route is Xewkija, ("the place where thistles grow") which is famous for its Rotunda church. This parish church, dedicated to St John the Baptist, was begun in 1951 and completed thirty years later, and was built by voluntary labour with the construction costs paid for by weekly contributions from the 3,500 villagers. The dome claims to be the third largest unsupported dome in the world, after St Paul's Cathedral in London and St Peter's in Rome. The church was constructed around and over its eighteenth century predecessor which was dismantled only after the new Rotunda was fit to be used towards the end of its construction time.

The bus rejoins the main road to Mgarr and so passes the site of the former Gozo Airport, and Gozo Heritage. The bus soon turns left off the main road to serve Ghajnsielem, which means "Salem's spring". The village's great neo-Gothic church was begun in 1922 and was finished over half a century later, in 1978, when the parish priest won the National Lottery and donated Lm10,000 for its new altar. The spire stands out amongst all the other church domes in the village.

Qala, the easternmost village on Gozo is the next village on the route; its name means 'sheltered harbour', and like Xaghra, boasts working windmills and a Baroque church.

Nadur is the final village to be served and has a very fitting name considering its lofty position with views across to Malta, for it means 'to keep guard'. Nadur is the most heavily populated of the villages on the island and is considered to be the village of emigrants. Its prosperity stems from the fact that its inhabitants have returned to Gozo after living abroad, especially in New York and New Zealand. Nadur was established as an autonomous parish on 28th April, 1688. The foundation stone of its Baroque church, dedicated to St Peter and St Paul Princes of the Apostles, was laid on 17th December, 1760 and the church was consecrated over a century later, in 1867. Also to be seen in Nadur is the Maritime Museum. Here there is a display of model sailing ships, naval emblems and uniforms covering 300 years of history, all of which was amassed by Kelinu Grima. The Kenuna Tower is another landmark, built by the British in 1848, on a spot where warning bonfires were lit during the time when the Knights of St John held sway. Its purpose was originally to serve as a telegraph link between Gozo and Malta. Indeed, Nadur's *Festa* day, 29th June each year, is called, in Maltese, l'Imnarja, 'the Feast of Illuminations', recalling the many bonfires that used to be lit on the eve and day of the *Festa*. From Nadur, the main road desends some sweeping bends and passes Gozo's Racecourse en route to Victoria.

But during July, August and September, Ramla Bay is served on various timings. The descent from Nadur to Ramla Bay provides wonderful views of the red-gold sands of the bay, fertile meadows, and on the hill opposite, Xaghra. The sand-dunes hereabouts contain flora and fauna almost unknown anywhere else on the islands, with marram grass, sand rest-harrow and sea spurge growing in profusion. Historically, Ramla Bay was perfect for invasion. So, in 1715 the Knights of Malta built two batteries on either side of the bay, and

FBY 054, a Bedford YMT with Plaxton Supreme III bodywork, is seen in the main square in Nadur. Pupils from the college in Victoria who had just travelled on the vehicle were already back in their home village by the time this picture was taken mid-morning.

also the "Vendôme redoubt" to provide further fortification. An underwater obstruction of 1730 vintage can still be seen offshore.

From Ramla Bay the bus continues its journey back to Victoria, climbing the hill on the other side of the valley, to Xaghra from where it returns to the terminus via route 64. An unusual feature of the Ramla Bay timetable is that the 1500 departure serves Xaghra too, and so buses can be seen bearing route numbers 64 and 42 on this timing.

Sunday and Public Holidays timings are at 0900 and 1100 from Victoria to Xewkija, Ghajnsielem, Qala, Nadur and return to Victoria.

The timings of school journeys from these villages are on display at the bus kiosk at Victoria:

Qala	0725	0750	0755
Tal Hniena	0730	0755	
Xewkija Roundabout	0734	0759	
Victoria	0739	0804	
St Anthony Street			0800
Ghajnsielem			0803
Garden Street			0805
Tal Hniena			0809
Victoria			0815

ROUTES 50/51 to SANNAT and MUNXAR

This is another of the shorter routes from Victoria to two villages served four times during the morning at 0715, 0900, 1000 and 1145, and then the final departure at 1715 which does not return to Victoria. The Monday to Saturday timings are such that the route is interworked with route 87 to Xlendi.

The route from Victoria is southwards along the direct main road to Sannat. This village boasts a church begun in 1718 with a fine altarpiece by a local artist, Stefano Errardi. Near the village are more prehistoric remains, the L-Imramma Temple and three dolmens. In Sannat the bus may travel up to Ta' Cenc to serve a large hotel complex built around a *palazzo* of the 17th century, which regularly features in lists of the world's top hotels. The precipitous cliffs nearby are over 120m high. Returning through Sannat, the bus bears left to travel along Triq il-Kbira passing the Lace House which is the former summer home of Princess Elizabeth and the Duke of Edinburgh and so heads towards Munxar, from where there is again easy access to the cliff-tops. The bus then returns directly to Victoria, with the whole journey taking about twenty minutes.

School journeys to these villages are displayed at the bus kiosk in Victoria:

Munxar Square	0730	0750	0810		
Sannat, Xelina Square	0800				
Ta' Cenc	0735				
Sannat Square	0735	0739	0755	0802	
Ta' Cawla	0738	0742	0758	0805	0810
Sannat School	0816				
Victoria	0743	0747	0803	0810	0815

The frontal appearance of **FBY 040** was severely altered a few years ago. It is seen leaving Victoria on route 51, despite what the route number indicates. It was pictured on the outskirts on Sannat en route to Munxar

ROUTES 64/65 to XAGHRA

This route runs directly to Xaghra, a hilltop village whose name refers to the wilderness of the hill before the area was built upon. On leaving the bus terminus at Victoria, the bus travels down the main road, bearing left at the first roundabout, along a dual-carriageway lined with oleanders (Gozo's national flower). A steep ascent involving two sharp (almost hairpin) bends brings the bus into Xaghra.

Many of the holiday-makers using this route come to Xaghra to visit the Ggantija Temples. These prehistoric temples date from 3600-3000 BC and are recognised as the oldest free-standing structures in the world. There are two temples with a common façade, but each has its own entrance. Local legend tells that the giantess Sansuna carried on her head the huge stones used for building the temples.

Nearby is the Ta' Kola Windmill which the bus will pass as it makes a short circuit of the roads near the Temples. The windmill was built in 1725 and was named after one of the first millers to work it. Still in working condition, it was used throughout the Second World War to provide wheat for the islanders.

There are thirteen timings of the route Monday to Saturday, starting at 0700, then every three-quarters of an hour from 0815 until 1200, 1315, then hourly from 1500 (1415, October to June) until 1900, this last timing not returning to Victoria. The 0700 and 0945 departures (as route 65) extend to Tan-Nazzarenu on the eastern edge of the village along the road leading to Ramla Bay. The 1500 departure proceeds to Ramla Bay from Xaghra, returning to Victoria via Nadur, as route 42.

On Sundays and Public Holidays departures from Victoria are at 1000 and 1200 to Xewkija roundabout, Xaghra and return but in July to September the 1315 is replaced by a journey at 1500 that continues beyond Xaghra to Ramla Bay, returning via Nadur to Victoria.

FBY 041, a Bedford YLQ with 10-metre Duple Dominant II bodywork approaches the Ggantija Temples in Xaghra where most of its passengers will be alighting.

ROUTE 87 to XLENDI

Lying on the south-west coast of Gozo, Xlendi is a typical fishing village, and in the summer, a favourite resort for holiday-makers. The route from Victoria passes through Fontana where there is a natural fresh-water spring (hence the name of the village), still in use today by local washerwomen. In the nineteenth century many of the fishermen at Xlendi lived in Fontana, and income from their catches went towards the building costs of the parish church dedicated to the Sacred Heart of Jesus.

On leaving Fontana the bus begins its descent to Xlendi and the terminus is amongst the palm-trees behind the waterfront. Cliffs border the inlet forming the village harbour, with Xlendi Tower, built in 1650 as a protection against naval attacks and capable of withstanding gunpowder shots, overlooking the harbour. The village, like Marsalforn, is a popular lunchtime destination for some of the coach-trips to the island.

Xlendi is served daily including Public Holidays throughout the year. During the high summer (July, August and September) there are twelve departures, at 0700, 0900, 1015, hourly from 1100 until 1500, 1615, 1730, 1830 and 2000. For the rest of the year the service boasts only seven timings, at 0930, 1115, 1230, 1400, 1515, 1615 and 1730.

FBY 003 is seen at the bus terminus in the main square of Xlendi in October 2007. The vehicle is one of the demonstrator buses imported to Malta during the mid-1990s for evaluation purposes.

MINIBUSES on MALTA and GOZO

Many minibuses were already in service in Malta and Gozo before the 1979 re-numbering. Only following the abolition of the twenty-five 'KIRI' midibus, minibus and midicoach permits were minibuses allowed to undertake any airport transfers. At that time there was no standard livery and vehicles were still regulated by the Unscheduled Bus Service. Between 1981 and 1987 government legislation imposed certain restrictions on the minibus fleet.

A new vehicle could not be purchased unless it was powered by a "York" diesel engine. So owners of any Ford Transit minibus with a Ford or Perkins engine had to fit a York engine before the vehicle could be sold on. Only twelve of the fourteen seats in the minibuses could be used, but this restriction infuriated the operators so much that they contested the ruling and obtained evidence from Ford that their vehicles were perfectly capable of transporting the maximum fourteen passengers. During these six years, as the number of red minuses increased, "new" operators who had previously been unemployed had to apply a white strip on the front wings of their vehicles to indicate this fact. In 1985 the 14-seater minibuses adopted the current red livery, whilst the rest of the garage-owned minibuses, which generally seated only ten passengers, adopted other colours, either cream or more usually white. Soon after the change of government in 1987, these six-year old restrictions were lifted and coincidentally the first Iveco replacements started to appear. The Minibus Co-operative Ltd was formed in 1988 and it was renamed Co-op Services Ltd eleven years later. Nowadays around eighty percent of the operators in the red minibus sector are members. The co-operative offers also road assistance, mobility services and minibus advertising.

This Asia AM 825, with registration IMY 339, is one of the Garden of Eden fleet and is seen waiting for the return of a troop of boy scouts who had been visiting the Playmobil Funpark in Hal Far in March 2009.

Currently there is a fleet of just over 400 red minibuses providing general tours, school services, works transport, private hire and other contract duties. Most carry a red livery with a white roof, although a few can be seen in a deeper, almost magenta, shade of red, or red combined with some other colour. Most are registered in the GMY, HMY and IMY series from 001 to 404. Unique in the fleet is Scarnif Saftran, HMY 152, which is now in full Cancu Supreme Travel colours rather than the red livery. It is the regular vehicle allocated to the Sliema local council service. Ford Transit minibuses of each marque remain in service. However Asia, Fiat, Isuzu, LDV Convoy, Mercedes Sprinter, Mitsubishi and Toyota Coaster vehicles are also in the fleet, along with some Mudan and King Long minibuses from China. The distinctive Scarnif Saftran vehicles are locally built midibuses. Since May 2009, new minibuses have received registrations in the new xPY series.

The Park and Ride scheme serving Valletta was introduced on 6th November 2006. Minibuses connect Freedom Square next to the remains of the old ruined Royal Opera House with the car parks at Blata l-Bajda. The red minibuses allocated to the service carry distinguishing Park & Ride transfers on their sides. Services operate every ten minutes between 0600 and 2100, after which it is every twenty minutes until 0100 on weekdays, and between 0600 and 0100 every twenty minutes at weekends and Bank Holidays.

On 6th February 2008, two more P&R routes from Blata l-Bajda were introduced. Operating every half-hour between 0930 and 1600 and then from 1800 and 2000 on weekdays, and between 0930 and 1300 on Saturdays, the routes pass through some of the narrow streets in the capital. Route A is the Blue route, operating via Old Bakery Street, Republic Street and St Elmo, whilst route B (Red) serves St Paul Street, St Christopher Street and Victoria Gate.

Unique amongst the minibus fleet is Cancu Supreme's Scarnif Saftran midibus, HMY 152, as it does not carry the traditional red livery of such vehicles. It is used primarily for the local town service in Sliema. It is seen along the Strand in Sliema awaiting its first journey of the day in September 2009.

Reg	Type	Code	Reg	Type	Code	Reg	Type	Code
FMY 001	Ford Transit	M14	GMY 070	Ford Transit	M14	HMY 139	Ford Transit	M14
GMY 001	Toyota Coaster	C18C	GMY 071	LDV Convoy	M17	HMY 140	Ford Transit	M14
LMY 002	Ford Transit	M14	GMY 072	Ford Transit	M14	HMY 141	Ford Transit	M14
GMY 003	Ford Transit	M14	GMY 073	Ford Transit	M14	HMY 142	Ford Transit	M14
GMY 004	Ford Transit	M14	GMY 074	Ford Transit	M14	HMY 143	Ford Transit	M14
GMY 005	Ford Transit	M14	GMY 075	Ford Transit	M14	HMY 144	Ford Transit	M14
GMY 006	Toyota Coaster	C18C	GMY 076	Ford Transit	M14	HMY 145	MB Sprinter	M18
LMY 007	Ford Transit	M14	GMY 077	Ford Transit	M14	HMY 146	Ford Transit	M14
GMY 008	Ford Transit	M14	GMY 078	Toyota Coaster	C18C	HMY 147	Ford Transit	M14
GMY 009	LDV Convoy	M16	GMY 079	LDV Convoy	M17	HMY 148	Ford Transit	M14
GMY 010	Ford Transit	M14	GMY 080	Ford Transit	M14	HMY 149	Ford Transit	M14
GMY 011	LDV Convoy	M17	GMY 081	Ford Transit	M14	HMY 150	Toyota Coaster	C18C
GMY 012	MB Sprinter	M18	GMY 082	Ford Transit	M14	HMY 151	Ford Transit	M14
GMY 013	Toyota Coaster	M18	GMY 083	Ford Transit	M14	HMY 152	Scarnif Saftran	C18F
GMY 014	LDV Convoy	M17	GMY 084	Ford Transit	M14	HMY 153	King Long	C18C
GMY 015	Ford Transit	M14	GMY 085	Ford Transit	M14	HMY 154	Ford Transit	M14
GMY 016	Ford Transit	M14	GMY 086	Ford Transit	M14	HMY 155	Ford Transit	M14
GMY 017	Ford Transit	M14	GMY 087	Ford Transit	M14	HMY 156	Ford Transit	M14
GMY 018	MB Sprinter	M18	GMY 088	Ford Transit	M14	HMY 157	Ford Transit	M14
GMY 019	Ford Transit	M14	GMY 089	Toyota Coaster	C18C	HMY 158	MB Sprinter	M18
GMY 020	Ford Transit	M14	GMY 090	Ford Transit	M14	HMY 159	Ford Transit	M14
GMY 021	Ford Transit	M14	GMY 091	Ford Transit	M14	HMY 160	Ford Transit	M14
GMY 022	Ford Transit	M14	GMY 092	Toyota Coaster	C18C	HMY 161	Kia Combi	C18C
GMY 023	Ford Transit	M14	GMY 093	Mudan	C18C	HMY 162	Scarnif Saftran	C18C
GMY 024	Ford Transit	M14	GMY 094	LDV Convoy	M17	HMY 163	Ford Transit	M14
GMY 025	Ford Transit	M14	GMY 095	Ford Transit	M14	HMY 164	Mitsubishi Rosa	C18C
GMY 026	Ford Transit	M14	GMY 096	MB Sprinter	M18	HMY 165	Ford Transit	M14
GMY 027	Ford Transit	M14	GMY 097	Ford Transit	M14	HMY 166	Ford Transit	M14
GMY 028	Ford Transit	M14	GMY 098	Ford Transit	M14	HMY 167	Ford Transit	M14
GMY 029	Scarnif Saftran	C18F	GMY 099	Ford Transit	M14	HMY 168	Ford Transit	M14
GMY 030	Ford Transit	M14	GMY 100	Ford Transit	M14	HMY 169	Ford Transit	M14
GMY 031	Toyota Coaster	C18C	GMY 101	Ford Transit	M14	HMY 170	Ford Transit	M14
GMY 032	Ford Transit	M14	GMY 102	MB Sprinter	M18	HMY 171	MB Sprinter	M18
GMY 033	Ford Transit	M14	GMY 103	Ford Transit	M14	HMY 172	Ford Transit	M14
GMY 034	Ford Transit	M14	GMY 104	Ford Transit	M14	HMY 173	Ford Transit	M14
GMY 035	MB Sprinter	M18	GMY 105	King Long	M18	HMY 174	Ford Transit	M14
GMY 036	Ford Transit	M14	GMY 106	Ford Transit	M14	HMY 175	Ford Transit	M14
GMY 037	Ford Transit	M14	GMY 107	Ford Transit	M14	HMY 176	Ford Transit	M14
GMY 038	Ford Transit	M14	GMY 108	MB Sprinter	M18	HMY 177	Ford Transit	M14
GMY 039	Ford Transit	M14	GMY 109	Ford Transit	M14	HMY 178	Ford Transit	M14
GMY 040	Ford Transit	M14	GMY 110	Ford Transit	M14	HMY 179	Ford Transit	M14
GMY 041	Ford Transit	M14	GMY 111	Ford Transit	M14	HMY 180	Ford Transit	M14
GMY 042	Ford Transit	M14	GMY 112	Ford Transit	M14	HMY 181	Ford Transit	M14
GMY 043	Mudan	C18C	GMY 113	Ford Transit	M14	HMY 182	Ford Transit	M14
GMY 044	Ford Transit	M14	GMY 114	Ford Transit	M14	HMY 183	MB Sprinter	M18
GMY 045	Fiat Ducato	M18	GMY 115w	Ford Transit	M14	HMY 184	Ford Transit	M14
GMY 046	Ford Transit	M14	GPY 012	Fiat Ducato	M14	HMY 185	King Long	C18C
GMY 047	MB Sprinter	M18	GMY 116	Ford Transit	M14	HMY 186	Ford Transit	M14
GMY 048	Ford Transit	M14	GMY 117	Ford Transit	M14	HMY 187	Ford Transit	M16
GMY 049	Ford Transit	M14	GMY 118	LDV Convoy	M17	HMY 188	Ford Transit	M14
GMY 050	MB Sprinter	M18	GMY 119	MB Sprinter	M18	HMY 189	Toyota Coaster	C18C
GMY 051	MB Sprinter	M18	GMY 120	Ford Transit	M14	HMY 190	Isuzu	M17
GMY 052	Ford Transit	M14	GMY 121	Ford Transit	M14	HMY 191	King Long	C18C
GMY 053	Ford Transit	M14	GMY 122	Ford Transit	M14	HMY 192	Ford Transit	M14
GMY 054	Ford Transit	M14	GMY 123	Asia AM825	M18	HMY 193	Ford Transit	M14
GMY 055	Ford Transit	M14	GMY 124	Ford Transit	M14	HMY 194	MB Sprinter	M18
GMY 056	Ford Transit	M14	GMY 125	Ford Transit	M14	HMY 195	Ford Transit	M14
GMY 057	Mitsubishi Rosa	C18C	GMY 126	LDV Convoy	M17	HMY 196	Ford Transit	M14
GMY 058	Ford Transit	M14	GMY 127	Ford Transit	M14	HMY 197	Ford Transit	M14
GMY 059	LDV Convoy	M17	GMY 128	Kia Combi	C18C	HMY 198	MB Sprinter	M18
GMY 060	Ford Transit	M14	GMY 129	Mudan	C18C	HMY 199	Ford Transit	M14
GMY 061	Ford Transit	M14	GMY 130	MB Sprinter	M18	HMY 200	Ford Transit	M14
GMY 062	Ford Transit	M14	HMY 131	MB Sprinter	M18	HMY 201	Ford Transit	M14
GMY 063	Ford Transit	M14	HMY 132	MB Sprinter	M18	HMY 202	Ford Transit	M14
GMY 064	Ford Transit	M14	HMY 133	MB Sprinter	M18	HMY 203	MB Sprinter	M18
GMY 065	Mudan	C18C	HMY 134	Ford Transit	M14	HMY 204	Mudan	C18C
GMY 066	LDV Convoy	M16	HMY 135	MB Sprinter	M18	HMY 205	Ford Transit	M14
GMY 067	Mudan	C18C	HMY 136	Ford Transit	M14	HMY 206	Ford Transit	M14
GMY 068	Scarnif Saftran	C18F	HMY 137	Ford Transit	M14	HMY 207	MB Sprinter	M18
GMY 069	Ford Transit	M14	HMY 138	MB Sprinter	M18	HMY 208	MB Sprinter	M18

Reg	Type	Code		Reg	Type	Code		Reg	Type	Code
HMY 209	Ford Transit	M14		IMY 278	Ford Transit	M14		IMY 347	Ford Transit	M14
HMY 210	Ford Transit	M14		IMY 279	Ford Transit	M14		IMY 348	Ford Transit	M14
HMY 211	MB Sprinter	M18		IMY 280	MB Sprinter	M18		IMY 349	Ford Transit	M14
HMY 212	MB Sprinter	M18		IMY 281	Ford Transit	M14		IMY 350	Ford Transit	M14
HMY 213	Mudan	M18		IMY 282	Mudan	C18C		IMY 351	Mudan	C18C
HMY 214	Ford Transit	M14		IMY 283	Ford Transit	M14		IMY 352	MB Sprinter	M18
HMY 215	Ford Transit	M14		IMY 284	Ford Transit	M14		IMY 353	MB Sprinter	M18
HMY 216	Ford Transit	M14		IMY 285	Ford Transit	M14		IMY 354	Ford Transit	M14
HMY 217	Ford Transit	M14		IMY 286	Ford Transit	M14		IMY 355	Ford Transit	M14
HMY 218	Ford Transit	M14		IMY 287	Toyota Coaster	C18C		IMY 356	Ford Transit	M14
HMY 219	Ford Transit	M14		IMY 288	Ford Transit	M14		IMY 357	Ford Transit	M14
HMY 220	Ford Transit	M14		IMY 289	Ford Transit	M14		IMY 358	Mudan	C18C
HMY 221	MB Sprinter	M18		IMY 290	Ford Transit	M14		IMY 359	Ford Transit	M14
HMY 222	MB Sprinter	M18		IMY 291	Fiat Ducato	M18		IMY 360	Kia Combi	C18C
HMY 223	Mudan	C18C		IMY 292	LDV Convoy	M17		IMY 361	Kia Combi	C18C
HMY 224	Ford Transit	M14		IMY 293	Ford Transit	M14		IMY 362	Ford Transit	M14
HMY 225	Ford Transit	M14		IMY 294	Toyota Coaster	C18C		IMY 363	Ford Transit	M14
HMY 226	Ford Transit	M14		IMY 295	Toyota Coaster	C18C		IMY 364	Ford Transit	M14
HMY 227	Ford Transit	M14		IMY 296	MB Sprinter	M18		IMY 365	Ford Transit	M14
HMY 228	LDV Convoy	M17		IMY 297	Ford Transit	M14		IMY 366	Ford Transit	M14
HMY 229	Ford Transit	M14		IMY 298	Ford Transit	M14		IMY 367	Ford Transit	M14
HMY 230	Ford Transit	M14		IMY 299	Ford Transit	M14		IMY 368	Ford Transit	M14
HMY 231	Ford Transit	M14		IMY 300	Ford Transit	M14		IMY 369	Scarnif Saftran	C18F
HMY 232	Ford Transit	M14		IMY 301	Ford Transit	M14		IMY 370	Ford Transit	M14
HMY 233	Ford Transit	M14		IMY 302	Ford Transit	M14		IMY 371	Toyota Coaster	C18C
HMY 234	Ford Transit	M14		IMY 303	Ford Transit	M14		IMY 372	Ford Transit	M14
HMY 235	Ford Transit	M14		IMY 304	LDV Convoy	M17		IMY 373	Ford Transit	M14
HMY 236	Ford Transit	M14		IMY 305	Ford Transit	M14		IMY 374	Ford Transit	M14
HMY 237	Ford Transit	M14		IMY 306	Ford Transit	M14		IMY 375	Ford Transit	M14
HMY 239	Ford Transit	M14		IMY 307	Ford Transit	M14		IMY 376	Ford Transit	M14
HMY 239	Ford Transit	M14		IMY 308	Kia Combi	M18		IMY 377	Ford Transit	M14
HMY 240	Ford Transit	M14		IMY 309	MB Sprinter	M18		IMY 378	Ford Transit	M14
HMY 241	Ford Transit	M14		IMY 310	Ford Transit	M14		IMY 379	Ford Transit	M14
HMY 242	Scarnif Saftran	C20C		IMY 311	Ford Transit	M14		IMY 380	Ford Transit	M14
HMY 243	Ford Transit	M14		IMY 312	Mudan	C18C		IMY 381	Ford Transit	M14
HMY 244	Ford Transit	M14		IMY 313	Mudan	C18C		IMY 382	Ford Transit	M14
HMY 245	Ford Transit	M14		IMY 314	MB Sprinter	M18		IMY 383	Ford Transit	M14
HMY 246	Ford Transit	M14		IMY 315	Ford Transit	M14		IMY 384	Ford Transit	M14
HMY 247	Ford Transit	M14		IMY 316	Fiat Ducato	M18		IMY 385	Ford Transit	M14
HMY 248	Ford Transit	M14		IMY 317	Ford Transit	M14		IMY 386	Ford Transit	M14
HMY 249	Ford Transit	M14		IMY 318	MB Sprinter	M18		IMY 387	Ford Transit	M14
HMY 250	Ford Transit	M14		IMY 319	Ford Transit	M14		IMY 388	Asia AM 825	C18C
HMY 251	LDV Convoy	M17		IMY 320	MB Sprinter	M18		IMY 389	Ford Transit	M14
HMY 252	Kia Combi	C18C		IMY 321	Ford Transit	M14		IMY 390	Ford Transit	M14
HMY 253	Ford Transit	M14		IMY 322	Ford Transit	M14		IMY 391	Toyota Coaster	C18C
HMY 254	Ford Transit	M14		IMY 323	MB Sprinter	M18		IMY 392	Ford Transit	M14
HMY 255	Ford Transit	M14		IMY 324	MB Sprinter	M18		IMY 393	Ford Transit	M14
HMY 256	Ford Transit	M14		IMY 325	Ford Transit	M14		IMY 394	Ford Transit	M14
HMY 257	Ford Transit	M14		IMY 326	Ford Transit	M14		IMY 395	Toyota Coaster	M18
HMY 258	Ford Transit	M14		IMY 327	Ford Transit	M14		IMY 396	Ford Transit	M14
HMY 259	Ford Transit	M14		IMY 328	Ford Transit	M14		IMY 397	Ford Transit	M14
HMY 260	LDV Convoy	M17		IMY 329	Ford Transit	M14		IMY 398	Ford Transit	M14
HMY 261	Toyota Coaster	C18C		IMY 330	Ford Transit	M14		IMY 399	Ford Transit	M14
HMY 262	Ford Transit	M14		IMY 331	Ford Transit	M14		IMY 400	Scarnif Saftran	C20F
IMY 263	Ford Transit	M14		IMY 332	LDV Convoy	M17		IMY 401	MB Sprinter	M18
IMY 264	Ford Transit	M14		IMY 333	Toyota Coaster	C18C		IMY 402	Ford Transit	M14
IMY 265	Ford Transit	M14		IMY 334	Ford Transit	M14		IMY 403	Ford Transit	M14
IMY 266	Ford Transit	M14		IMY 335	Ford Transit	M14		IMY 404	Scarnif Saftran	C18
IMY 267	Ford Transit	M14		IMY 336	Kia Combi	C18C		GPY 012	Fiat Ducato	M18
IMY 268	Ford Transit	M14		IMY 337	Ford Transit	M14		GPY 013	Ford Transit	M18
IMY 269	Ford Transit	M14		IMY 338	Ford Transit	M14		HPY 013	LDV Convoy	M17
IMY 270	Ford Transit	M14		IMY 339	Asia AM825	C18C		HPY 015	LDV Maxus	M17
IMY 271	Ford Transit	M14		IMY 340	Ford Transit	M14		JPY 012	MB Sprinter	M18
IMY 272	Ford Transit	M14		IMY 341	Ford Transit	M14		JPY 013	MB Sprinter	M18
IMY 273	Ford Transit	M14		IMY 342	Ford Transit	M14		KPY 012	Ford Transit	M18
IMY 274	Ford Transit	M14		IMY 343	Ford Transit	M14		KPY 015	LDV Maxus	M17
IMY 275	Ford Transit	M14		IMY 344	Ford Transit	M14		MPY 002	Peugeot	
IMY 276	Ford Transit	M14		IMY 345	Ford Transit	M14		MPY 007	Ford Transit	M18
IMY 277	Ford Transit	M14		IMY 346	Fiat Ducato	M18				

OPEN-TOP BUSES

The history of open-top vehicles on Malta dates back to 1993, when both Garden of Eden and Cancu Supreme imported three open-top double deckers from Britain. During the ensuing years neither company was permitted to provide any regular public service for the deckers, either on tours or on private hire, as governmental and police restrictions prevented the vehicles from operating legally on the road. It was only after each general election during these years that the vehicles were ever seen on the roads, with one or other of the two companies providing transport for the winning political party, with the new Prime Minister and his cabinet on board touring the island to celebrate their victory, with Garden of Eden vehicles for the Labour Party and Cancu Supreme vehicles when the Nationalist Party was victorious.

So, for about thirteen years after their arrival on the island, the deckers were stored inside at the companies' depots, eventually encased in layers of polythene as some form of protection against the elements and to avoid unnecessary damage. It was not until 2006, when the ATP invited interested parties to submit applications to operate two open-top tourist routes which would be known as the Northern and Southern Routes, starting from Sliema Ferries, that there was any likelihood of the double deckers being seen in service. The ten-year contract was awarded to Cancu Supreme in December 2006 and the original open-toppers, the trio of Fleetlines, were the first vehicles to be equipped for the tours, with multi-language commentary systems, seat-belts and hand-rails being fitted. Further imports followed bringing the fleet up to fifteen in total.

The management at Garden of Eden waited a long time before receiving permission to operate their fleet of Open-Top buses. Here is COY 013 in its City Sightseeing livery biding time in Marsaxlokk on the South Tour of the island early one afternoon in September 2009.

Cancu Supreme's COY 011 is seen near the Corinthia Marine Hotel in St George's Bay whilst working the Northern tour, and hence its blue livery. Because of a road traffic accident on the main road in St Andrew's, all vehicles had been diverted around the St George's peninsula, thus covering a section of route 66 from Valletta.

The Southern (red) route began operation on 2nd July 2007, originally without Sunday services, which were subsequently introduced in November 2008. The Northern (blue) route commenced early the following year, in January 2008.

Both of Cancu's Open-Top routes have been adapted to meet the demands of local councils and road conditions, but have now settled down to provide an extensive daily service. The main routes operate hourly between 0900 and 1500 from Sliema Ferries on weekdays and between 1000 and 1300 on Sundays and Bank Holidays. Feeder services from hotels in Bugibba, Qawra, Salina and St Julian's are provided so as to link up with the earlier departures from the main terminus at Sliema Ferries. So, buses leave the Santana Hotel in Bugibba at 0815 (but not on Sundays and Public Holidays), 0915 and 1015 every day, and arrive at Sliema Ferries half an hour later.

The Northern Route is timed to take two and quarter hours and covers the following route from Sliema Ferries: Manoel Island, Ta'Xbiex, Msida, Valletta Porta Reale bus terminus, Hamrun, San Anton Gardens, Mdina Glass, Ta'Qali Crafts Village, Mosta, Mdina, Rabat, Dwejra Lines, Mgarr, Golden Bay, Bugibba (various hotels), St Julian's, Balluta Bay and return to Sliema Ferries.

The Southern Route lasts slightly longer – two hours twenty five minutes – and follows the same route as its Northern partner as far as Valletta Castle Gate terminus, and then continues to many of capital's principal sights starting with Upper Barracca Gardens, then Valletta harbour, War Museum, Fort St Elmo, the Mediterranean Conference Centre, Fort St Angelo, and Valletta Waterfront. Leaving the capital, the bus then goes to Marsa, the Hypogeum and Temples at Tarxien, Senglea Point, Vittoriosa Waterfront, Fort Rinella, Marsaxlokk, San Lucijan Tower, Ghar Dalam Caves, the Blue Grotto, Hagar Qim, the Limestone Heritage, Malta Falconry Centre, Montekristo Estates at Luqa and then non-stop return via Santa Venera to Sliema Ferries.

The immediate success of these daytime tours encouraged Cancu to introduce night tours on Monday, 12th May 2008. Connecting departures from Bugibba are provided at 1800, 1900 and 2000 to Sliema Ferries, from where the main tours start forty-five minutes later. These night tours operated daily from 16th March until 15th November during 2009.

From Bugibba, the route covers the Northern Coast Road via Paceville and St Julian's, Sliema Ferries, Ta'Xbiex, Msida, Valletta, Castille, Fort St Elmo, the Mediterranean Conference Centre, the Aqueducts near Santa Venera, Balzan, Lija and Attard, Naxxar, Mosta, Rabat and Mdina, and then returns to the Bugibba hotels, Paceville and Sliema Ferries.

On 1st February 2010, Cancu introduced the first open-top tour on Gozo. A connecting service from hotels in Valletta and the Sliema area and one from hotels in Bugibba, Qawra and Mellieha are timed to arrive at Cirkewwa at 0845 in time for the 0900 ferry to Mgarr on Gozo. The open-top route is based on Mgarr and operates hourly from 0930 until 1430 every day except Christmas Day. The anti-clockwise tour lasts two hours and takes in the Ggantija Temples, Xaghra, Calypso's Cave, Marsalforn, Victoria, Ta' Pinu, San Lawrenz Crafts Village, Dwejra, Fontana and Xlendi. Some of the route covers roads not otherwise served by Gozo's route buses – to reach Calypso's Cave and between Xlendi, Sannat and Xewkija on the return leg of the trip to Mgarr.

Cancu Supreme's other tour is the one to the Three Cities which departs from Sliema Strand at 1030 and 1430. One of Cancu's "old-timers" (LCY 001, 002 or 004) normally operates the service. Once again a detailed commentary is provided giving passengers the opportunity to hear about the history of Valletta and the Three Cities and the two Great Sieges of 1565 and 1942. The route to the Three Cities passes through Gzira, Ta'Xbiex, Msida, Pieta, Marsa and Paola, and then tours the squares and narrow streets of Cospicua, Senglea and Vittoriosa.

The directors of Garden of Eden continued their campaign to introduce their own Open-Top services during the early months of the Cancu operations and eventually they were granted licences to provide a wholly different type of tour. In conjunction with City Sightseeing the company provides all-day tours which allow passengers to spend time at various locations before rejoining their bus, rather than waiting for the next departure to arrive, as happens with the Cancu tours. These City Sightseeing Tours were introduced in May 2009.

The North Tour operates Monday to Saturday and offers pick-up facilities from hotels in Cirkewwa, Mellieha, (both of these points have to be pre-booked), Bugibba and St Julian's. Departure from Sliema Ferries is at 1015 from where the route goes via Gzira, Birkirkara Bypass and Balzan to the first destination at Mosta Dome where there is a twenty minute stop. Then there is a forty minute visit to Ta' Qali Crafts Village, twenty minutes at Rabat and one hour at Mdina, with the return journey passing through Mgarr, Golden Bay for a photo stop, St Paul's Bay, Qawra and the Northern Coast Road back to Sliema Ferries.

The South Tour operates each day of the week with similar hotel pick-ups as the North Tour and a departure time of 1030 from Sliema Ferries. The route passes through Valletta to the first stop for half an hour at the Three Cities, then another half-hour visit, this time to the Tarxien Temples, a lunch-time visit and slide-show about Maltese wine-making at the Montekristo Unique Experience, followed by the Blue Grotto for three quarters of an hour and then a similar stay at Marsaxlokk. The return journey goes via St Lucijan Tower, the quarries at Zurrieq, Marsa and return to Sliema Ferries.

The other tours on offer are a Sunday Tour to the markets at Valletta and Marsaxlokk and the Montekristo Unique Experience; a Fireworks Night each Saturday (from 23rd May until 26th September during 2009); Folklore Night at the Limestone Heritage at Siggiewi on Fridays (from 22nd May until 30th October during 2009); a Malta by Night tour from Bugibba, via Sliema, Valletta, Floriana and Santa Venera to Mdina. Time is allowed for a walk through the "Silent City", before the return journey via Mosta and St Paul's Bay. This tour operated on Mondays to Thursdays from 18th May until 30th September during 2009, and for the winter months from 1st October 2009 and on into 2010 on Mondays to

Oasis Tours was another company which imported a double-decker from Britain and then converted it to open-top configuration. This Metrobus was latterly in the West Midlands Travel fleet. However, with the introduction of the Halcrow recommendations for transport reform on the island, it seems increasingly unlikely that the bus will ever be seen in service. POG592Y was photographed outside the bar, restaurant and take-away of Oasis Tours in the Hal Farrug Industrial Estate near Luqa.

Wednesdays; the Malta Festa Night Tour operated each Sunday from 24th May until 27th September during 2009.

Other companies on the island, notably Oasis Tours, had also obtained their own open-top vehicles from Britain — or rather, had imported double-deckers which were then converted to OT formation, but none of them has received the authority to operate such routes and it seems increasingly unlikely, as the months go by, that these other OT buses will ever be seen in service.

The tourist routes now operated by Cancu and Garden of Eden are not as ground-breaking as might be thought. They are in fact successors of two "Visit Malta Tourist Bus Services" provided by the ADT during the mid-2000s. These routes, numbered 505 and 506, lasted for two complete summers and then were summarily withdrawn with only forty-eight hours advance notice one week before the Easter weekend in 2005. One stipulation of these routes was that only the older traditional routes buses were to be rostered for the daily departures in an effort to attract further patronage.

Route 505 was the North (red) Route and to a certain extent foreshadowed Cancu's Open-top services later in the decade. It departed from Sliema Ferries at 0915, 1000, 1200, 1245, 1445 and 1530, thus two vehicles were required each day to provide the service – one providing the first, third and fifth departures, the other the second, fourth and sixth timings. The buses would travel along the Northern Coast Road to Qawra, then to St Paul's Bay, Mosta Dome, Mdina, Buskett Gardens, Dingli, Magdalena Chapel on Dingli Cliffs, Rabat, Mtarfa bypass, Mosta, Naxxar, San Gwann, Paceville and return to Sliema Ferries. The round trip was timetabled to take two hours thirteen minutes with the major departure times en route being advertised as "fixed".

Route 506 was the South (blue) Route and again foreshadowed the later Cancu Open-Top service. It too departed from Sliema Ferries, but at different times from the North Route, at 0900, 0930, 1215, 1300, 1530 and 1615, with identical arrangements for two vehicles to operate the service each day. This route also served Paceville from Sliema

Ferries, then San Anton Gardens, Siggiewi, Hagar Qim, Wied iz-Zurrieq, Bir Miftuh Chapel, Marsaxlokk, Kalkara, Cottonera, Paceville and return to Sliema Ferries and the round trip took two hours fifty minutes.

A trip on these routes cost Lm 2. Tickets were available from the bus driver and were valid for travel on both routes each day, as well as any scheduled service up to 2300.

The final tourist attraction started on Saturday 9th August 2008. This "Historical and Cultural Train Tour" is operated by Melita Trains using a Dotto Train built in Italy, which consists of a diesel-powered engine hauling three carriages with seats for a maximum of sixty passengers. A commentary in various languages is available and the system is GPS operated. The departure point is the Roman Villa on the outskirts of Mdina, close to the lay-over area for the Unscheduled buses. During the half-hour tour the train goes along the narrow streets of Rabat past the catacombs, through the residential estate of Ghar Barka, along the main road through Mtarfa, before joining the main Mtarfa Bypass where there are panoramic views over to Mosta and Mdina. Turning left off the bypass, the train serves the terminus of the former railway in Malta at Notabile and skirts the edge of the "Silent City", before plunging uphill through a tunnel back to the terminus. During the summer of 2009 departures were at 1000, 1130, 1230, 1400, then hourly from 1530 until 2130, and during the autumn and on into the winter of 2010 hourly between 1000 and 1600.

				New	To Malta	In Service	
COY 001	Leyland Fleetline FE30AGR	Eastern Coach Works	043/31F	9/78	5/93	6/07 Cancu Supreme "Calypso"
COY 002	Leyland Fleetline FE30AGR	Eastern Coach Works	043/31F	8/78	5/93	6/07 Cancu Supreme "Joy"
COY 003	Leyland Fleetline FE30AGR	Eastern Coach Works	043/31F	8/78	5/93	6/07 Cancu Supreme "Three Hills"
COY 004	AEC Routemaster 3R2RH	Park Royal	041/31F	2/65		6/07 Cancu Supreme
COY 005	Volvo Citybus B10M-50	Alexander RV	047/37F	10/89		10/07 Cancu Supreme
COY 006	Leyland Titan TNLXB2RR	Leyland	045/32F	8/84		1/07 Cancu Supreme
COY 007	Leyland Titan TNLXB2RR	Leyland	044/32F	7/83		1/08 Cancu Supreme
COY 008	Leyland Titan TNLXB2RR	Leyland	044/31F	7/83		2/08 Cancu Supreme
COY 009	Leyland Titan TNLXB2RR	Leyland	044/32F	3/83		2/08 Cancu Supreme
COY 010	Leyland Titan TNLXB2RR	Leyland	044/32F	7/83		4/08 Cancu Supreme
COY 011	Volvo Citybus B10M-50	Alexander RV	047/37F	11/89		5/08 Cancu Supreme
COY 012	Volvo Citybus B10M-50	Alexander RV	047/37F	10/89		8/08 Cancu Supreme
COY 013	Leyland Fleetline FE30AGR	Eastern Coach Works	043/31F	4/77	7/93	12/08 Garden of Eden
COY 014	Leyland Fleetline FE30AGR	Eastern Coach Works	043/31F	4/77	6/93	12/08 Garden of Eden
COY 015	Daimler CVG6	Roe	033/26F	10/68	7/93	12/08 Garden of Eden
LPY 015	Scania N230 UD	Optare Visionaire	0N51/30F	11/09		1/10 Cancu Supreme "Faith"
LPY 016	Scania N230 UD	Optare Visionaire	0N51/30F	11/09		1/10 Cancu Supreme "Hope"
LPY 017	Scania N230 UD	Optare Visionaire	0N51/30F	11/09		1/10 Cancu Supreme "Charity"
xPY nnn	Volvo B7L	Ayats	0N55/24F	6/04		(-/10) Garden of Eden
xPY nnn	Volvo B7L	Ayats	0N55/24F	6/04		(-/10) Garden of Eden

COY 001, UMR192T	Thamesdown 192
COY 002, UMR191T	Thamesdown 191
COY 003, UMR193T	Thamesdown 193
COY 004, FPT588C	Big Bus Co RMF588; Blue Triangle RMO2118; Go-Ahead 2118
COY 005 G301OGE	FirstBus 30371; Strathclyde PTE AH53
COY 006, A56THX	Big Bus Co CRM1056, London Transport T1056
COY 007, OHV811Y	Big Bus Co EM2000; London Transport T811
COY 008, OHV792Y	Big Bus Co LB4792; London Transport T792
COY 009, OHV720Y	Big Bus Co CMB1720; London Transport T720
COY 010, OHV808Y	Big Bus Co CLB4808, London Transport T808
COY 011, G688PNS	FirstBus 31566; Strathclyde PTE AH60
COY 012, G294OGE	FirstBus 30366; Strathclyde PTE, AH46
COY 013, OHR189R	Thamesdown 189
COY 014, OHR188R	Thamesdown 188
COY 015, JVV266G	McKeever, Leicester; Northampton Corporation 266
LPY 015	New
LPY 016	New
LPY 017	New
u/r, EU04CUW	Ensignbus 382
u/r EU04CZR	Guide Friday (Ensignbus), 384

POLICE BUS FLEET

All ten vehicles currently comprising the Malta Police Buses fleet formerly operated in England.

In the recent past the fleet contained Freight Rover Sherpas once owned by National Welsh. The livery of the Bedford VAS5 buses is dark blue to below the windows and white above, whilst the Iveco buses are white with a blue band. The fleet is garaged at Police Headquarters in Floriana.

GVP 159	Bedford VAS5	Wadham Stringer	B25F	9/81	by 6/92
GVP 160	Bedford VAS5	Wadham Stringer	B27F	10/81	by 6/92
GVP 161	Bedford VAS5	Wadham Stringer	B24F	11/81	by 6/92
GVP 163	Bedford VAS5	Wadham Stringer	B22F	5/81	by 6/92
GVP 494	Iveco 59-12	Marshall C31	B25F	10/95	2/03
GVP 495	Iveco 59-12	Marshall C31	B25F	8/94	2/03
GVP 497	Iveco 59-12	Marshall C31	B25F	12/95	2/03
GVP 498	Iveco 59-12	Marshall C31	B29F	9/94	2/03
GVP 499	Iveco 59-12	Marshall C31	B25F	9/94	2/03
GVP 500	Iveco 59-12	Marshall C31	B25F	12/95	2/03

GVP 159, M-0762, BGJ 830X	London Borough of Croydon, J005
GVP 160, M-0763, BGJ 837X	London Borough of Croydon, A039
GVP 161, M-0764, BGJ 840X	London Borough of Croydon, A053
GVP 163, M-1744, HUW 602W	London Borough of Tower Hamlets, 56804
GVP 494, N740 AVW	Arriva The Shires, 2360
GVP 495, M725 UTW	Arriva The Shires, 2345
GVP 497, N741 AVW	Arriva The Shires, 2361
GVP 498, M151 RBH	Arriva The Shires, 2091
GVP 499, M721 UTW	Arriva The Shires, 2341
GVP 500, N744 AVW	Arriva The Shires, 2364

One of the Police Bedford VAS5 vehicles, GVP 160, is seen at the Police Headquarters in Floriana.

EDUCATION DEPARTMENT FLEET

The Ministry of Education fleet has reduced to six operational vehicles in recent years. The age profile of the buses has led to an increasing dependence on the older M-registered buses still in their spray-green livery providing spares. The former parking area in Floriana closed in the mid-2000s. The fleet is maintained at the garage and repair facility in Pembroke. The six buses still in use can be often found parked at a school in Hamrun during school hours. The trio of Swifts are in a blue and cream livery, GVH 133 is purple and cream and the two Ford coaches are white with purple bands.

GVH 130	AEC Swift 4MP2R	Marshall	B46F	4/70	4/82
GVH 132	AEC Swift 4MP2R	Park Royal	B43D	10/70	2/88
GVH 133	AEC Swift 4MP2R	Park Royal	B43D	10/70	4/82
GVH 134	Ford R1114	Plaxton Supreme	C53F	12/78	5/92
GVH 135	Ford R1114	Plaxton Supreme	C53F	4/79	5/92
GVH 137	AEC Swift 4MP2R	Park Royal	B43D	1/71	4/82
GVH 140	AEC Swift 4MP2R	Park Royal	B46D	10/70	4/82
GVH 141	AEC Swift 4MP2R	Park Royal	B42D	1/71	4/82
M-1511	AEC Swift 4MP2R	MCW	B—D	1/71	4/82
M-1512	AEC Swift 4MP2R	Park Royal	B—D	10/70	4/82
M-1516	AEC Swift 4MP2R	Marshall	B—D	10/70	4/82
M-1519	AEC Swift 4MP2R	MCW	B—D	8/71	4/82

GVH 130, M-1531, AML35H	London Transport, SM35
GVH 132, M-0557, EGN260J	London Transport, SMS266
GVH 133, M-0786, EGN266J	London Transport, SMS266
GVH 134, M-1591, CBM294T	Prosser, Birchgrove; Bruton, London SW 11
GVH 135, M-1581, XWN9T	Prosser, Birchgrove; Jenkins, Skewen
GVH 137, M-1525, EGN337J	London Transport, SMS337
GVH 140, M-1522, EGN259J	London Transport, SMS259
GVH 141, M-1521, EGN356J	London Transport, SMS356
M-1511, JGF736K	London Transport, SMS736
M-1512, EGN278J	London Transport, SMS278
M-1516, EGN180J	London Transport, SMS180
M-1519, EGN602J	London Transport, SMS602

One of the two Fords with Plaxton bodywork in the Education fleet, GVH135, descends Constitution Hill in Mosta, close to the premises of Paramount Garage in September 2009.

AIRPORT VEHICLES

Since January 2006 there have been two companies providing the transfer of passengers to the terminal at Malta International Airport. The Globeground Malta vehicles are primarily used by low-cost airlines which now operate to the islands.

Air Malta

Fleet Number	Chassis	Bodywork	Seating	New
1 AB 326	Neoplan Airliner	Neoplan	B10FD	by 5/93
2 AB 327	Neoplan Airliner	Neoplan	B10FD	by 5/93
3 AB 328	Neoplan Airliner	Neoplan	B10FD	by 5/93
4 AB 329	Neoplan Airliner	Neoplan	B10FD	by 5/93
5 AB 384	Neoplan Airliner	Neoplan	B10FD	by 3/02
6 AB 385	Neoplan Airliner	Neoplan	B10FD	by 3/02
7 AB 402	Contrac Cobus 3000	Cobus	B10SD	by 8/03
8 AB 403	Contrac Cobus 3000	Cobus	B10SD	by 8/03
9 AB 440	Neoplan Airliner N9122	Neoplan	B10SD	by 3/08
10 AB 441	Neoplan Airliner N9122	Neoplan	B10SD	by 3/08
11 AB 453	Neoplan Airliner N9122	Neoplan	B10SD	by 3/08

Globeground

1 u/r	Contrac Cobus 3000	Cobus	B10SD	by 1/06
2 u/r	Contrac Cobus 3000	Cobus	B10SD	by 1/06
3 u/r	Contrac Cobus 3000	Cobus	B10SD	by 1/06

Three withdrawn vehicles can still be seen within the airport grounds. These were originally in service with Deutsche Bundesbahn before being exported to Malta:

9 AB 22	MAN	MAN	B44D	registration AMG 043
10 AB 23	MAN	MAN	B44D	
11 AB 24	MAN	MAN	B44D	registration AMG 044

EBY 537 is a Ford Thames ET7 with Micallef bodywork and is one of the three remaining normal-control Route Buses still in daily operations on Malta. It is photographed at the terminus of route 91 in its home town of Qormi.

ISBN 9781904875581 © Published by *British Bus Publishing Ltd*, February 2010

**British Bus Publishing Ltd, 16 St Margaret's Drive, Telford, TF1 3PH
Telephone: 01952 255669**

web; www.britishbuspublishing.co.uk - e-mail: sales@britishbuspublishing.co.uk